The Mathematics of Photosynthesis
and Productivity

EXPERIMENTAL BOTANY

An International Series of Monographs

CONSULTING EDITORS

J. F. Sutcliffe

School of Biological Sciences,
University of Sussex,
Falmer, Brighton, UK

and

J. Cronshaw

Department of Biological Sciences,
University of California,
Santa Barbara, California 93106, USA

A complete list of titles in this series is given at the end of the book.

The Mathematics of Photosynthesis and Productivity

D.A. Charles-Edwards

Biomathematics Department
Glasshouse Crops Research Institute
Littlehampton, UK

Present address: Division of Tropical
Crops and Pastures, CSIRO, St. Lucia,
Queensland, Australia

1981

ACADEMIC PRESS
A Subsidiary of Harcourt Brace Jovanovich, Publishers
London   New York   Toronto   Sydney   San Francisco

ACADEMIC PRESS INC. (LONDON) LTD.
24/28 Oval Road,
London NW1 7DX

United States Edition published by
ACADEMIC PRESS INC.
111 Fifth Avenue
New York, New York 10003

British Library Cataloguing in Publication data
Charles-Edwards, D. A.
    The mathematics of photosynthesis and productivity.
    – (Experimental botany series)
    1. Photosynthesis – Mathematical models
    I. Title   II. Series
    581.1'3304207'2   QK882

    ISBN 0-12-170580-3
    LCCCN 81-66387

Typeset by Gilbert Composing Services, Leighton Buzzard, Bedfordshire

Preface

In 1967 John Warren Wilson, the newly appointed Head of the Plant Physiology Department at the Glasshouse Crops Research Institute, outlined his proposals for the development of research in the Department. He argued, both cogently and persuasively, that a programme of mathematical modelling should be associated with the experimental research programme. As a result of these proposals John Thornley was, in 1969, appointed as Head of the Biometrics Department and given the responsibility to "construct mathematical models of crop and plant processes". I was appointed to the Biometrics Department in 1972, having previously held an ARC post-Doctoral Fellowship at the Welsh Plant Breeding Station, where I worked under John Cooper on the measurement and analysis of genetic variation in leaf photosynthetic activity. Not unnaturally, my previous experience determined the nature of my main modelling work at GCRI. This monograph is an unashamedly personal and partial account of a large part of my researches during the past ten years. It has been my good fortune to have been directly associated with either the gathering or primary analysis of most of the experimental data that I have reported here.

My thanks are due to colleagues at GCRI for their help and encouragement over the past years. In particular, I would like to thank John Thornley, John Ludwig, Basil Acock, Alun Rees and David Hand for their sustained enthusiasm and help.

In September, 1979, at the end of a year spent in Brisbane as Visiting Research Fellow with the CSIRO Division of Tropical Crops and Pastures, I was quite badly hurt in a car accident. Merv Ludlow and Myles Fisher, with whom I had been working, unselfishly and unstintingly cared for my family, and, when I left hospital, encouraged me to write this monograph. On my return to GCRI Derek Rudd-Jones, Director of the Institute, and John Thornley added their encouragement and practical help. I hope this monograph will be some small repayment to them all for their support and enthusiasm throughout the long, and often frustrating, months of my recovery.

The monograph is not intended to be any sort of definitive text on modelling and photosynthesis. I have sought to show how quite simple mathematics, mostly simple algebra, can be used to formalise ideas, and aid the analysis and ordering of experimental data. I have no particular attachment of any one of the models described. They all represent very simple descriptions of quite complex natural phenomena and there are more precise and comprehensive descriptions of many of the phenomena. However, it is their separate simplicities that allow them to be integrated into an understanding of the growth and productivity of field crops.

Finally, I would like to thank Doreen Crane and Maurice Bone for their tremendous help in the physical preparation of the manuscript and illustrations.

January 1981   D.A. C-A.

For Merv and Myles

Acknowledgements

The author wishes to thank the publishers of publications listed below for permission to reproduce some of the Figures used in this book.

Figs. 3.3, 3.4 and 4.2: *Annals of Botany* **41**, 1977.

Figs. 3.8, 3.9 and 3.10: *Annals of Botany* **40**, 1976.

Fig. 4.3: *Annals of Botany* **44**, 1979.

Figs. 4.4, 4.5, 4.6 and 4.7: *Annals of Botany* **46**, 1980.

Fig. 2.10: *Journal of Experimental Botany* **25**, 1974.

Figs. 3.1 and 3.2: *Journal of Experimental Botany* **27**, 1976.

Fig. 3.6: *Journal of Experimental Botany* **29**, 1978.

Figs. 3.11, 3.12, 3.13 and 3.14: *Annals of Applied Biology* **90**, 1978.

Fig. 2.4: "Environmental and Biological Control of Photosynthesis",
  R. Marcelle, Junk, 1975.

Contents

4. Photosynthesis and Productivity

1. Introduction

1.1. INTRODUCTION

It is not intended that this book should be regarded as any sort of definitive text on either the modelling or the descriptive physiology of leaf or crop photosynthesis. It has been written in the hope that it will illustrate the thesis that by the judicious application of quite simple mathematics to problems in a well-researched area of plant physiology we can improve our understanding of both the interdependencies of the many different plant processes and their separate dependencies on the plant's external environment. It is therefore relevant to start by attempting to define the phrase 'mechanistic mathematical model', and then to examine how such a model might contribute to our investigations into the nature and functioning of plant processes.

A mechanistic mathematical model is a mathematical description of a particular plant process, which attempts to describe the process quantitatively in terms of its constituent chemical and physical reactions. It is a set of mathematical equations; equations which separately provide formal statements of the assumptions that have been made in setting up the particular hypothesis about the plant process. Clearly, if the assumptions can be written down formally and unambiguously in the language of mathematics we are more likely to be able to derive a formal mathematical statement of the hypothesis itself. It is then rational to suppose that with a formal mathematical statement of the hypothesis we would be better able to examine its consequences and to explore whether or not it could account for the observed behaviour of the plant, or plant process. If the model is unable to account for the observed behaviour it might still help us to identify those areas where our knowledge of the plant, or plant process, is inadequate, and where further experimental studies are needed. A satisfactory model of a plant process may also help us to integrate an understanding of that process into an understanding of the performance of the intact plant, or crop.

It needs to be emphasized that the techniques of mathematical modelling have been used for many years in the physical sciences, and any student of physics or chemistry would accept them unquestioningly as a necessary, an integral, and indeed a very ordinary part of their chosen science. Any general textbook in either of these sciences abounds with practical examples of the

1

important and essential role that mathematical models have played in the development of our understanding of the physical and chemical interactions of substances. For example, both the theoretical and experimental study of the kinetics of chemical reactions would have made little progress without the use of modelling techniques.

In direct contrast to the physical sciences, the plant sciences and their associated agricultural sciences, have made little use of these techniques. This has undoubtedly been due to their traditional roles and applications as descriptive sciences rather than analytical sciences. Their traditional roles are not surprising. There is a great diversity of both plant type and plant growth environment. Each needs to be catalogued, and then the two need to be related, one to the other. It is appropriate that these needs have caused the plant sciences to provide a considerable impetus to the development of descriptive mathematics, and in particular to the mathematics of statistics. However, with the development of the more analytical branches of the plant sciences there is an increasing need to establish a framework of hypothesis within which we can order our knowledge of the processes of plant development and growth. It is in this context that mathematical models become useful and important tools to the furtherance of our understanding of the behaviour of plants and their responses to changes in their environments.

1.2. TESTING A HYPOTHESIS

A mechanistic mathematical model of a plant process is simply a formal mathematical statement of a hypothesis that has been advanced about the process. Like all hypotheses, it is not too difficult to conceive ways of disproving a model's validity. However, it is impossible, in any absolute sense, to prove its validity. Proof relies on the subjective assessment of its robustness and success in bringing order to apparently disordered observations. Broadly, there are two approaches that can be used to help test a model's validity. The first is to use the mathematics to predict the quantitative behaviour of the plant process being modelled, and then to compare the predictions with the observed behaviour of the process. If the prediction is numerically at variance with the experimentally observed behaviour of the process the model may, in the strictest sense, be deemed wrong. However, the approach demands accurate numerical values for the model's parameters to be known, and the prediction may be wrong simply because the wrong numerical information was fed into the model at the outset. The second, less rigorous, approach, is to compare the qualitative behaviour of the model with the observed behaviour of the process. If the qualitative behaviour of the model is similar to the behaviour of the real process, it would be of interest to know what numerical values of the model's parameters would provide a good quantitative description of the plant process. This leads to the important procedure known as 'fitting'. Since the procedure is an important technique in the modelling approach it is worthwhile examining it in more detail.

Let us suppose that we solve a particular model to obtain the set of dependent variables

$$Y_1, Y_2, Y_3, \ldots, Y_{n-1}, Y_n,$$

and let us further suppose that this set compares exactly with an analogous set of experimental data, which we can denote by

$$y_1, y_2, y_3, \ldots, y_{n-1}, y_n.$$

We can now compare the numerical values of the 'predicted' data with the real data. We can define the residual sum of squares, R, by

$$R = \sum_{i=1}^{n} (y_i - Y_i)^2, \tag{1.1}$$

and it will provide a measure of the goodness-of-fit of the model. Now, the predicted values $Y_1, \ldots, Y_n$ will depend on the parameters contained in the mathematical equations which constitute the model. The 'best fit' of the model can be sought by adjusting some, or all, of the numerical values of the model's parameters so that the residual sum of squares, R, is a minimum. For example, if the model has only one parameter, which we can denote by P, the residual sum of squares will be a minimum if

$$\frac{\mathrm{d}R}{\mathrm{d}P} = 0 \quad \text{and} \quad \frac{\mathrm{d}^2 R}{\mathrm{d}P^2} > 0. \tag{1.2}$$

An important corollary of eqn (1.2) is that P will be most closely defined if $\mathrm{d}^2 R/\mathrm{d}P^2$ is large, that is if R changes rapidly with changes in P. This is demonstrated in Fig. 1.1, where the broken curve illustrates the type of relationship between R and P for a model with a sensitive parameter and the solid line represents a model with an insensitive parameter.

If the model is deterministic the predicted values $Y_1, \ldots, Y_n$ are obtained without an associated probability distribution. The experimental data set $y_1, \ldots, y_n$ will, however, be subject to error. This will put a lower limit to the value of R that can be attained by adjusting the model's parameters. We can divide the residual sum of squares into two parts, so that

$$R = R_e + R_f, \tag{1.3}$$

where R_e is the component due to error in the experimental data and R_f the component due to lack of fit of the model. If the model contains m parameters,

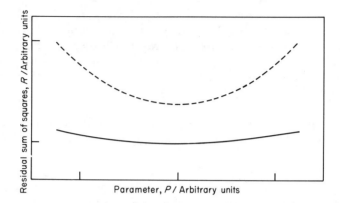

FIG. 1.1. Models with sensitive and insensitive parameters. In the broken curve the residual, R, is sensitive to the numerical value of the parameter P. This contrasts with the solid curve where the numerical value of P has far less effect upon R.

R_e will have a value of

$$R_e = (n - m)\sigma^2, \qquad (1.4)$$

where σ^2 is the error variance. The parameters $P_1, \ldots, P_m$ will only affect the component R_f, and if R_e is much larger than R_f the residual sum of squares, R, might prove to be very insensitive to the parameter values. If the data are replicated the component R_e might be known. In this case an upper limit to R_e is given by the minimum value of R obtained by adjusting the model's parameters.

Where appropriate, the residual sum of squares can be obtained for 'predicted' and experimental data sets that have been weighted in some way. For example, it may be more appropriate to write

$$R = \sum_{i=1}^{n} [\ln(y_i) - \ln(Y_i)]^2, \qquad (1.5)$$

and minimize this logarithmically weighted residual sum of squares.

1.3. PERSPECTIVES

A nineteenth-century chemist wrote:

> When we consider that the many thousand tribes of vegetables are not only all formed from a few simple substances, but that they all enjoy the same sun, vegetate in the same medium, and are supplied with the same nutrient, we cannot but be struck with the rich economy of Nature, and

are almost induced to doubt the evidence of those senses with which the God of Nature has furnished us.

If we are to retain our objectivity it is important to put the essential role of photosynthesis to the continuance of life on this planet in perspective with its primary role as an important determinant of plant growth. A distinctive property of plants, and all other living organisms, is that they are ordered and capable of creating order from their less ordered surroundings. This property of orderliness reflects the fact that they all depend on some external source of energy for their existence. In the language of the thermodynamicist, they are all 'open systems'. On the global scale this energy source is the sun or, more correctly, that portion of the sun's radiant energy that reaches the earth's surface. It is because the process of photosynthesis is almost unique as a means by which this energy is converted to a stable, storable, chemical form available for use by living organisms that photosynthesis achieves its pre-eminence over other biological processes. The photosynthetic process is effected by plants; indeed the ability of living organisms to undertake the process provides us with a simple, unambiguous definition of the word 'plant'. Plants use the products of the photochemical reactions of photosynthesis to mediate the synthesis of organic compounds from gaseous carbon dioxide derived from the earth's atmosphere. Although photosynthesis is only one of a number of processes by which plants acquire material from their environment, in this case the material is carbon, the process is of fundamental importance to all other life forms. All other living organisms gain access to the sun's energy by consuming, either directly or indirectly, plants or plant parts.

In the context of plant growth the process of photosynthesis, or carbon acquisition, is no more or no less important than any other process by which the plant assimilates essential materials from its environment. For example, a plant will not grow unless it is able to assimilate nitrogen and phosphorus from the soil. Photosynthesis cannot, therefore, be studied in isolation from these other processes. The physiological study of photosynthesis as an important determinant of plant growth needs to be directed at defining and analysing the inter-relationships between all the primary assimilating processes. The techniques of mathematical modelling are a logical and necessary aid to such a study.

Whilst many parts of the plant are capable of photosynthesis, the photosynthetic process is usually associated with its leaves. For most plants the leaves represent the major light intercepting surface, and thereby account for the greatest part of the plant's photosynthetic activity. If we are intending to study the role of the photosynthetic process as an important determinant of plant and/ or crop growth and productivity, it is therefore logical and sensible that we should first attempt to understand and describe the response of leaf photosynthesis to changes in the leaf's environment. Our understanding should attempt to encompass both the instantaneous response of the leaf to changes in its environment, and its longer-term adaptive responses, associated with changes in the plant's growth environment. If we have an understanding of these responses we will then be in a position to integrate this understanding into our analysis of the role of photosynthesis as a determinant of plant and/or crop productivity. We would hope that our subsequent analysis at the plant or crop levels would allow us to interpret the behaviour at these levels in terms of the identified behaviour of individual leaves. If this proves to be so, we are then better able to predict the effects of changes in the plant or crop environment on its photosynthetic performance and subsequent productivity.

1.4. PHOTOSYNTHETIC PATHWAYS

Common to all green plants are the sequence of chemical reactions which are known as the Calvin cycle. In this sequence of reactions a key role is played by the enzyme ribulose bisphosphate carboxylase (RubP carboxylase). This enzyme catalyses the reaction between the acceptor molecule ribulose-1,5-bisphosphate

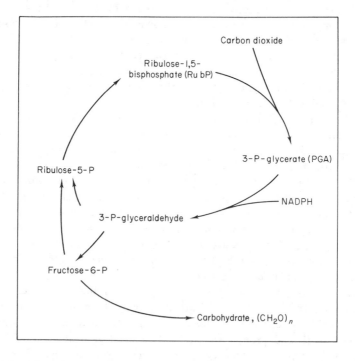

FIG. 1.2. Schematic representation of the processes of the Calvin cycle. (NADPH= nicotinamide adenine dinucleotide phosphate (reduced form)).

(RubP) and carbon dioxide. The carbon dioxide is derived from the gaseous carbon dioxide in the bulk air surrounding the leaf. The product of the reaction catalysed by RubP carboxylase is 3-phosphoglycerate (PGA). This product, 3-phosphoglycerate, is then metabolized by the further reactions of the Calvin cycle. These reactions lead to the formation of a carbohydrate moiety, (CH_2O), and the regeneration of RubP. The synthesis of the carbohydrate moiety is, of course, the essential feature of the Calvin cycle and, more generally, the process of photosynthesis. A schematic representation of the reactions of the Calvin cycle is illustrated in Fig. 1.2. The Calvin cycle represents the 'dark' reactions of the photosynthetic process. It is sufficient to note here that the energy required to operate the cycle is derived from the photochemical reactions, or 'light' reactions, of photosynthesis. It is these 'light' reactions which give rise to the nicotinamide adenine dinucleotide (reduced form) shown in the scheme. The detailed mechanisms of both the 'dark' and 'light' reactions are dealt with at some length in most good biochemical textbooks.

The enzyme RubP carboxylase appears to have another very important property. It is able to mediate the oxidation of ribulose bisphosphate to phosphoglycolate. The compound phosphoglycolate is then readily metabolized by the plant to the amino acid glycine with the associated production of carbon dioxide. It has been suggested that this oxygenase activity of RubP carboxylase arises because an intermediate anion of RubP, formed in association with the enzyme during photosynthesis, is inherently susceptible to oxidation by oxygen. Whatever the exact mechanism, it appears that the oxygenase activity of the photosynthetic enzyme RubP carboxylase is a crucial, and controlling, factor in the process of photorespiration.

The word photorespiration simply means the light dependent oxidation of organic metabolites to carbon dioxide, which is then respired. The process of photorespiration is clearly the exact opposite of the process of photosynthesis. Photosynthesis is simply defined as the light dependent synthesis of organic metabolites from gaseous carbon dioxide.

It is known from studies of the gas exchange of intact leaves that in many plants oxygen appears to act as a competitive inhibitor to the uptake of carbon dioxide during leaf photosynthesis. This observation could be readily explained if we were to postulate that the same enzyme, RubP carboxylase, mediated both the processes of photosynthesis and photorespiration. However, in some plants an increase in the concentration of gaseous oxygen in the bulk air surrounding the leaf appears to have little effect on the photosynthetic process when it is observed at the leaf level. Either this group of plants has evolved quite different reaction sequences for the photochemically activated fixation of carbon dioxide or they have evolved some means by which they are able to suppress the oxygenase activity of the enzyme RubP carboxylase.

When these observations of the different gas exchange patterns of leaves are combined with biochemical and anatomical studies we are able to distinguish two main groups in the higher plants. In the first, the C_3 group, carbon dioxide derived directly from the bulk air surrounding the leaf is fixed by the leaf in the production of the three carbon compound PGA. This carbon fixation process is mediated by the enzyme RubP carboxylase. In the second, the C_4 group, the carbon dioxide derived from the bulk air surrounding the leaf is initially fixed in the production of the four carbon compound oxaloacetate (OAA). The leaves of these plants also have a particular cellular structure, known as the Kranz anatomy. The production of OAA is mediated by the enzyme phosphoenol-pyruvate carboxylase (PEP carboxylase). Subsequently, a product derived from OAA is de-carboxylated, and the carbon dioxide produced is refixed in the production of PGA by the normal reactions of the Calvin cycle. Associated with these different reaction pathways are the structural differences between C_3 and C_4 leaves. In the leaves of the C_3 plants the processes of the Calvin cycle occur in the tissues of the palisade and spongy mesophyll cells which are distributed uniformly across the length and breadth of the leaf. In the C_4 plants the initial fixation of carbon dioxide by PEP carboxylase occurs in the tissues of the mesophyll cells. However, the processes of the Calvin cycle are primarily confined to the bundle sheath cells, an agglomeration of chlorenchyma cells

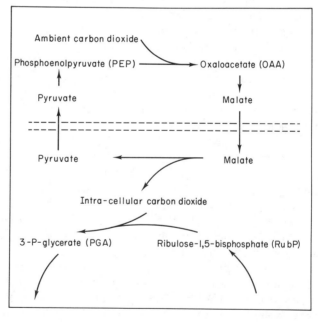

FIG. 1.3. Schematic representation of the initial processes of carbon dioxide fixation in C_4 plants.

which surround each of the vascular bundles. In the C_3 plants, therefore, the initial process of carbon dioxide fixation and PGA production are the same, and occur at the same reaction site. In contrast, in the C_4 plants the processes are different, and occur at spatially separate reaction sites. A simple scheme for the processes in C_4 plants is illustrated in Fig. 1.3. We can now add the observation that it is the C_3 plants which exhibit the apparent competitive inhibition of photosynthesis by oxygen, at the leaf level, and it is the C_4 plants which appear to be· insensitive to changes in the oxygen concentration in the bulk air surrounding the leaf.

It is now only a small step to hypothesize that the initial reaction of carbon dioxide fixation in C_4 plants, the reaction mediated by the enzyme PEP carboxylase, serves as a 'carbon dioxide pump' in these plants. That is, by spatially separating the processes by which carbon dioxide is initially fixed from the bulk air from the processes of the Calvin cycle, the effective concentration of carbon dioxide available to the Calvin cycle reactions is increased. This would allow the carbon dioxide to compete more effectively with the oxygen, derived from the atmosphere, at the site of the RubP carboxylase/ oxygenase activity.

1.5. LEAF PHOTOSYNTHESIS

Leaves. exhibit a variety of morphological and metabolic responses to changes in their physical environment, and the causes of these different responses can sometimes be inferred from their associated time constants. For example, whereas the instantaneous response of leaf photosynthesis to a change in the light flux density incident on the leaf surface reflects the kinetic properties of the existing photosynthetic apparatus, differences in the absolute rates of leaf photosynthesis arising from long-term differences in the light environment during the growth of the leaf are more likely to reflect differences in the size, activity or display of that apparatus. Clearly, if we are able to characterize the kinetic properties of the photosynthetic apparatus by analysis of the instantaneous responses of that apparatus to changes in the leaf's environment we will be in a better position to identify and analyse the longer term adaptive response of the leaf. Further, if these functions are derived from an understanding of the physical and chemical reactions involved in leaf photosynthesis we may be able to analyse the differences between leaves at a considerable level of mechanistic detail.

There are numerous mathematical descriptions (models) of leaf photosynthesis. They range from simple empirical descriptions of the phenomenon to detailed mechanistic analyses of the biochemical and physical reactions of the photosynthetic process. In choosing any one of these models to analyse a particular data set, two considerations are of great importance. First, what is the objective of the analysis of that data set? Second, what level of resolution is available with the particular data set? For plant physiologists the objective in making such an analysis will probably be to help identify and study one of the following:

(a) Differences in leaf photosynthetic activity between material of contrasting genetic origin.
(b) Adaptive responses of leaves to contrasting plant growth environments.
(c) Ontogenetic changes in leaf photosynthetic activity.
(d) The main components of plant photosynthesis, and thence plant dry matter production.

Whilst for studies at the leaf level, therefore, a suitable model would preferably have some basis in the known reaction mechanisms of the photosynthetic process, at the plant or crop levels a simpler, more empirical model might be appropriate.

It has been suggested that none of the mechanistic models of leaf photosynthesis should be taken very seriously since they generally assume homogeneity of the characteristics and local environments of the individual chloroplasts within the leaf. It has been argued that an important part of the modelling problem is to allow for the possible heterogeneity of the intra-leaf environment. Whilst these arguments carry some force, the technical problems associated with obtaining experimental information on the intra-leaf environment are formidable, and it seems unlikely that progress in this particular area of research will be very rapid. Similarly, whilst some research workers have looked at the mathematical consequences to leaf photosynthesis models of different assumptions about the global kinetics of the Calvin cycle, it seems unlikely that the predicted mathematical differences could be effectively distinguished at the leaf level. Indeed, even if the heterogeneity of the intra-leaf environment, or the stochiometric detail of the biochemical reactions, were explicitly taken into account, the inherent variability of measurements at the leaf level would more than likely prevent their unambiguous resolution.

Perhaps the most widely used mathematical description of leaf photosynthesis is the one shown in eqn (1.6). This equation relates the rate of leaf photosynthesis per unit leaf area, F, to the light flux density incident on the leaf's surface, I, and the ambient carbon dioxide concentration, C. The equation is

$$F = \alpha I \tau^* C / (\alpha I + \tau^* C) - R_d \qquad (1.6)$$

where R_d is the rate of dark respiration by the leaf, α is the leaf's photochemical efficiency and τ^* its conductance to carbon dioxide uptake. The main deficiencies of eqn (1.6) are that it does not describe the effects of changes in ambient oxygen concentration on the rate of photosynthesis by leaves of C_3 plants, nor does it explicitly deal with the processes by which carbon dioxide moves from the bulk air surrounding the leaf to the photosynthetic sites within the leaf. However it is readily extended to deal with this latter deficiency. Recently, a number of models have been reported which seek to remedy these deficiencies. Although they differ in detail, their mathematical properties are very similar. The mathematical effects of changing ambient oxygen concentrations are based on the hypothesis that RubP carboxylase mediates both the processes of photosynthesis and, through its oxygenase activity, photorespiration.

Equation (1.6) describes a rectangular hyperbola for the response of the leaf's net photosynthesis rate to changes in the leaf's light and carbon dioxide environments. The parameter α is the initial slope of the light response curve, and the product αI the asymptotic value of $F + R_D$ at saturating ambient carbon dioxide concentrations. Similarly, τ^* is the initial slope of the carbon dioxide response curve, and τ^*C the asymptotic value of $F + R_D$ at saturating light flux densities.

The reciprocal of τ^*, known as the leaf 'resistance', is commonly used in leaf photosynthesis studies, and results from the use of electrical analogues to describe photosynthesis. Historically, models of leaf photosynthesis have tended to emphasize the physical restraints imposed on the process by the diffusion of carbon dioxide from the air to the photosynthetic sites. As a result of this emphasis it became common for plant physiologists to use electrical analogues to describe the photosynthetic process. There is a patent similarity between the physical restraints to gas diffusion and the restraints to electron flow in an electrical circuit. However, unless the biochemical processes of photosynthesis can be shown to respond to changes in the leaf's environment, in an analogous manner to the diffusive processes, the extension of electrical analogues to include the biochemical processes can be misleading.

It is perhaps pertinent to remind ourselves that a mathematical model is simply a formal statement of the assumptions that have been made about a particular process. Even if the details of the process are obscure, we may have good reason to expect a mathematical model of the process to show the right sort of qualitative behaviour if the general concept that it seeks to formalize is correct. If the single most important feature of the photosynthetic and photorespiratory processes is the competition between carbon dioxide and oxygen for the same acceptor molecule (ribulose-1,5-bisphosphate), and their reactions are catalysed by the same enzyme (RubP carboxylase/oxygenase), we might expect this competition to dominate the mathematical description of the two processes. A pseudo-mechanistic mathematical model based on this competition would most likely provide a good qualitative description of the process. Indeed, the model might provide a good quantitative description when 'fitted' to experimental data, although the numerical estimates of the model's parameters might not allow the fine detail of the biochemical and physical processes to be unequivocally resolved. In the following chapters I will develop and explore mathematical models of leaf and canopy photosynthesis based on the thesis that photosynthesis and photorespiration are regulated by the same enzyme (RubP carboxylase/oxygenase). I hope to demonstrate that the use of such models can bring order to a good deal of otherwise disordered experimental information.

1.6. SUGGESTED FURTHER READING

Section 1.1

Charles-Edwards, D. A. and Thornley, J. H. M. (1974). *Span* **17**, 57-59.

Section 1.2

Thornley, J. H. M. (1976). *In* 'Mathematical Models in Plant Physiology', Ch. 1. Academic Press, London and New York.
Bell, C. J. (1981). *In* 'Mathematics and Plant Physiology', Ch. 17. (D. A. Rose and D. A. Charles-Edwards, eds). Academic Press, London and New York.

Section 1.4

Chollet, R. and Ogren, W. L. (1975). *Bot. Rev.* **41**, 137–179.
Lorimer, G. H. and Andrews, T. J. (1973). *Nature, Lond.* **243**, 359–360.
Mahler, M. R. and Cordes, E. H. (1966). *In* 'Biological Chemistry', Ch. 11. Harper and Row, New York, Evanston and London.
Devlin, R. M. (1975). *In* 'Plant Physiology', Chs. 10–12. D. Van Nostrand Co., New York.

Section 1.5

Thornley, J. H. M. (1976). *In* 'Mathematical Models in Plant Physiology', Ch. 4. Academic Press, London and New York.

2. Leaf Photosynthesis

2.1. INTRODUCTION

This chapter is concerned with both the instantaneous and adaptive responses of leaf photosynthesis to changes in the leaf's environment. The first part deals with the construction, test and application of a simple pseudo-mechanistic mathematical model for leaf photosynthesis. The model is used to analyse the instantaneous response of leaf photosynthesis to changes in the ambient carbon dioxide and oxygen concentrations of the leaf and the light flux density incident on the leaf. The effects of changes in leaf temperature and water status are also examined. The model is then extended to examine possible effects of changes in the leaf's anatomy and structure on its photosynthetic performance. This extension allows the model to be used to analyse some of the adaptive responses of leaves, and the differences between leaves grown under the same environmental conditions but derived from different genetic material. Finally, having constructed and tested the model and used it to analyse the behaviour of real leaves its contribution to the physiological understanding of leaf photosynthesis will be discussed.

Sections 2.2 and 2.3 deal with the mathematical descriptions of the biochemical and physical processes involved in leaf photosynthesis. When we attempt the mathematical description of our physiological hypotheses about these processes we need to make a number of assumptions. The assumptions are usually made to enable the mathematics to be solved, or to retain some degree of simplicity, but it is important for us to be aware of the limitations that they may place on our mathematical description. The separate analyses of the biochemical and physical processes are then brought together in Section 2.4 to provide a pseudo-mechanistic model for the leaf net photosynthesis.

Some properties of the model are examined in Section 2.5, and its qualitative behaviour is compared with the observed photosynthetic behaviour of real leaves. Values of the model's parameters are obtained by 'fitting' the model to experimental data. The 'fitting' procedure has been discussed in Section 1.2. Briefly, values of the model's parameters are chosen so as to minimize the sum of squares of the differences between the observed data values and the corresponding values predicted by the model. A 'good fit' does not validate the model. It simply means that with the particular numerical values of its parameters the model provides a good quantitative description of the experimental data. The critical test of a model is whether or not it provides a consistent and ordered description of the leaf's photosynthetic behaviour, creating order from apparently disordered observations of the leaf's behaviour in contrasting environments.

Temperature is unique amongst the environmental variables in that it affects all of the individual reactions which together constitute the process of net photosynthesis. In general it affects the rates of chemical reactions more than the rates of physical reactions (reactions such as gaseous diffusion). Whilst useful information can be obtained from a study of the effects of temperature on the rate of leaf net photosynthesis, it may not be clear from the analysis of this type of data which of the individual reactions is the most susceptible to temperature changes. It is more logical to investigate the effects of temperature on the parameters which, through the model, are believed to characterize the kinetic properties of the process. It is then to be hoped that the analysis of this information will reveal which reaction(s) is the most sensitive to changes in leaf temperature. However, such a programme of research and analysis would require considerable experimental effort. In Section 2.6 some aspects of the potential effects of changes in the leaf's temperature on its instantaneous rate of net photosynthesis are examined.

Temperature is also a major factor in determining the efficiency with which a plant uses water. We can define the term 'water use efficiency' as the ratio of the rates of water loss by the leaves and their rates of net photosynthesis. Leaves lose water through the process of transpiration, which is essentially the evaporative loss of water. The leaf tissues within the epidermis are presumed to have water-saturated cell surfaces, and the inter-cellular air spaces are assumed to be saturated with water vapour, at the leaf temperature. Water vapour diffuses, via the stomates, from these air spaces into the bulk air surrounding the leaf. As the leaf's temperature rises so does the saturation vapour pressure within the inter-cellular air spaces. If the cross-sectional area of the stomates remains constant and the concentration of water vapour in the bulk air does not change, an increase in leaf temperature will lead to an increase in the transpiration rate. When the leaf transpiration rate exceeds the rate of leaf water uptake the leaf will begin to experience a water deficit. Plants respond to increasing leaf water deficits in several different ways. Some plants have specialized tissues which enable them to change both the inclination and orientation of the leaf. When leaves of these plants experience severe enough water deficits they alter both inclination and orientation to lie 'edge on' to the sun. This minimizes the radiation flux densities incident on the leaf, and thereby the leaf's total energy receipt and hence its temperature. These and other plants are also able to reduce leaf transpiration rates by closing their stomates as their leaf water deficits increase. In closing their stomates these plants also reduce their rates of net photosynthesis, and this effect is examined in Section 2.7.

Leaves are able to adapt to longer term changes in their environments. The environment experienced by the leaf during its period of growth affects its structural development. In Section 2.8 the model for leaf net photosynthesis is extended to take into account some possible effects of leaf structure in determining the rates of the physical processes of gaseous diffusion between the bulk air and the photosynthetic sites. This analysis is then used in Section 2.9 to examine the adaptive response of leaf photosynthesis to different light and temperature environments during growth. Nutritional status of the plant also affects leaf photosynthetic activity. This is examined in Section 2.10, together with the genetic variability observed in leaf photosynthetic characteristics.

Having constructed a mathematical model for leaf photosynthesis and then having compared some of its properties with the behaviour of real leaves it is pertinent to ask what contribution the modelling exercise has made to our physiological understanding of the process. The role and value of the modelling approach are discussed in Section 2.11. Finally the main symbols used in the mathematics are defined and their dimensions given.

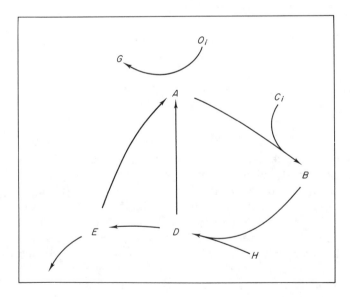

FIG. 2.1. Schematic representation of the processes of photosynthesis and photorespiration. A, B and D represent intermediates of the Calvin cycle, E and G are products of the two processes and H is a reducing agent produced by the 'light' reactions of photosynthesis. O_i and C_i represent oxygen and carbon dioxide available at the photosynthetic site.

2.2. THE BIOCHEMICAL PROCESSES OF PHOTOSYNTHESIS

The most distinctive feature of the biochemical processes of the 'dark' reactions of photosynthesis is their cyclic nature. A simple representation of these reactions is shown in Fig. 2.1, and this scheme is usefully compared with the more explicit scheme for the Calvin cycle shown in Fig. 1.2. Following the arguments of Section 1.4, the process of photorespiration can be treated as the irreversible removal of ribulose bisphosphate, by oxidation, from the Calvin cycle. That would correspond to the removal of the intermediate A, shown in Fig. 2.1.

There is likely to be a simple analogy between the kinetic behaviour of the scheme shown in Fig. 2.1 and the behaviour of a bi-substrate enzyme reaction, with carbon dioxide (C) and the reducing agent H being the two substrates. The concentrations of enzyme and enzyme/substrate or enzyme/product complexes would need to be replaced by the concentrations of the intermediates of the photosynthetic cycle. The removal of the intermediate A, through oxidation, would be treated as the competitive interaction of oxygen and carbon dioxide for the same acceptor molecule. The scheme shown in Fig. 2.1, with both the oxidation and carboxylation of the acceptor molecule A, can be re-written as the following set of 'chemical' equations:

$$A + C_i \xrightarrow{\ j_1\ } B$$
$$A + O_i \xrightarrow{\ j_2\ } G$$
$$B + H \xrightarrow{\ j_3\ } D$$
$$D \xrightarrow{\ j_4\ } A + E$$
$$E \xrightarrow{\ j_5\ } A.$$

The symbols j_1–j_5 denote rate constants; A, B and D are the concentrations of intermediates of the photosynthetic cycle; G and E are products of photorespiration and photosynthesis. The scheme contains three independent variables: C_i and O_i are the concentrations of carbon dioxide and oxygen at the photosynthetic site and H is the concentration of reducing agent produced by the 'light' reactions of photosynthesis. The scheme also includes the assumption that the acceptor molecule A can be re-synthesized from the products of photosynthesis (E). We can write the rates of gross photosynthesis, P_G (that is gross carbon fixation) and photorespiration, R_L as

$$P_G = j_1 A C_i \tag{2.1}$$

$$R_L = j_2 A O_i. \tag{2.2}$$

It is important to note that since we are dealing with concentrations, or perhaps densities, of reactants in the photosynthetic system both P_G and R_L represent rates expressed on the basis of the unit volume of the system. We can assume that under steady-state conditions the concentrations of the cycle intermediates $(A, B$ and $D)$ will not change, and we can write

$$\frac{dA}{dt} = \frac{dB}{dt} = \frac{dD}{dt} = 0. \tag{2.3}$$

If we further assume that the total concentration of both the cycle intermediates $(A, B$ and $D)$ and the products of photosynthesis (E) remain constant, we can add

$$A + B + D + E = \sigma = \text{constant.} \tag{2.4}$$

The set of differential equations describing the kinetic behaviour of the reaction scheme during the steady state can now be written as

$$\frac{dA}{dt} = 0 = j_4 D + j_5 E - j_1 A C_i - j_2 A O_i$$

$$\frac{dB}{dt} = 0 = j_1 A C_i - j_3 B H$$

$$\frac{dD}{dt} = 0 = j_3 B H - j_4 D.$$

These equations lead directly to the relationships

$$B = j_1 A C_i / j_3 H \tag{2.5}$$

$$D = j_1 A C_i / j_4 \tag{2.6}$$

$$E = j_2 A O_i / j_5, \tag{2.7}$$

which, when substituted into eqn (2.4), yield the identity

$$A(j_3 j_4 j_5 H + j_1 j_4 j_5 C_i + j_3 H(j_1 j_5 C_i + j_2 j_4 O_i))/j_3 j_4 j_5 H = \sigma,$$

which rearranged will give

$$A = j_3 H \sigma / (j_3 H + j_1 C_i + j_3 H(j_1 C_i / j_4 + j_2 O_i / j_5)). \tag{2.8}$$

If the concentration of the reducing agent H is proportional to that portion of the light flux density incident on the leaf's surface which is absorbed by the leaf, we can also write that

$$H = aQ \tag{2.9}$$

where Q is the light energy absorbed per unit volume of photosynthetic system and a is a proportionality constant. Although eqns (2.4)–(2.8) are exact mathematical descriptions of the steady-state behaviour of the reaction scheme shown in Fig. 2.1, they are no more than empirical descriptions of the real photosynthetic process. It is useful to re-emphasize this empiricism by re-parameterizing them. If we define the five new parameters α_m, τ, β, δ and γ by

$$\alpha_m = aj_3\sigma$$

$$\tau = j_1\sigma$$

$$\beta = j_2\sigma$$

$$\delta = 1/j_4\sigma$$

$$\gamma = 1/j_5\sigma$$

we can show, by substitution of eqns (2.8) and (2.9) into (2.1) and (2.2), that

$$P_G = \alpha_m Q\tau C_i/(\alpha_m Q + \tau C_i + \alpha_m Q(\delta\tau C_i + \gamma\beta O_i)) \qquad (2.10)$$

and

$$R_L = \alpha_m Q\beta O_i/(\alpha_m Q + \tau C_i + \alpha_m Q(\delta\tau C_i + \gamma\beta O_i)). \qquad (2.11)$$

Equations (2.10) and (2.11) relate the rates of gross photosynthesis and photo-respiration to the three independent variables Q, C_i and O_i.

Equations (2.10) and (2.11) have some interesting properties. If either the absorbed light energy, Q, is large, or the terms in the denominators containing δ and γ are sufficiently large, they reduce to

$$P_G = \tau C_i/(1 + \delta\tau C_i + \gamma\beta O_i) \quad \text{or} \quad P_G = \tau C_i/(\delta\tau C_i + \gamma\beta O_i)$$

and

$$R_L = \beta O_i/(1 + \delta\tau C_i + \gamma\beta O_i) \quad \text{or} \quad R_L = \beta O_i/(\delta\tau C_i + \gamma\beta O_i).$$

These new expressions describe competitive inhibition between carbon dioxide and oxygen for their separate reactions with the acceptor molecule A. It should be noted that the term $\gamma\beta O_i$ arises because the photosynthetic product E has been included in the assumption that the total concentration of intermediates and products of photosynthesis remains constant (eqn (2.4)). If the assumption is further restricted to exclude the concentration E (i.e. we assume that $A + B + D$ = constant), the oxygen concentration, O_i, and then the term $\gamma\beta O_i$, will not appear in the denominators of eqns (2.10) and (2.11). In any case, the oxygen concentration, O_i, only appears in association with the absorbed light energy, Q. If the reaction between the acceptor molecule, A, and the carbon dioxide is assumed to be reversible, the oxygen concentration, O_i, will appear alone in the denominator.

Equations (2.10) and (2.11) are strictly empirical, based on an 'educated guess' at the global kinetics of the processes of photosynthesis and photo-respiration. We need to add one other important relationship to them. If we define P as the rate of net photosynthesis and R_D as the rate of 'dark' respiration (oxidative phosphorylation), both expressed per unit volume, simple logic requires that

$$P = P_G - R_L - R_D. \qquad (2.12)$$

Equation (2.12) records the logical statement that the net flux of carbon dioxide through the photosynthetic apparatus is the difference between the rate at which it is consumed by the process of photosynthesis and the rates at which it is produced by the processes of dark respiration and photorespiration.

The five parameters α_m, τ, β, γ and δ can be usefully re-defined in terms of the physiological properties of the photosynthetic apparatus. The initial slope of the response of gross photosynthesis to changing light levels can be shown to be given as

$$\left[\frac{dP_G}{dQ}\right]_{Q=0} = \alpha_m, \qquad (2.13)$$

which defines α_m as the light utilization efficiency of gross photosynthesis. Similarly we can define τ and β by

$$\left[\frac{dP_G}{dC_i}\right]_{C_i=0} = \tau/(1 + \gamma\beta O_i) \qquad (2.14)$$

and

$$\left[\frac{dR_L}{dO_i}\right]_{O_i=0} = \beta/(1 + \delta\tau C_i). \qquad (2.15)$$

Equation (2.14) allows us to define τ as the maximum carboxylation efficiency for gross photosynthesis in the absence of oxygen, and eqn (2.15) allows us to define β as the maximum oxygenation efficiency for photorespiration in the absence of carbon dioxide. At high light levels and high carbon dioxide concentrations, but at low oxygen concentrations, we can write

$$P_G(Q, C_i \to \infty, O_i \to 0) = 1/\delta, \qquad (2.16)$$

which defines δ as the reciprocal of the maximum potential rate of gross photosynthesis. Similarly at high light levels and high oxygen concentrations, but at low carbon dioxide concentrations,

$$R_L(Q, O_i \to \infty, C_i \to 0) = 1/\gamma, \qquad (2.17)$$

which defines γ as the reciprocal of the maximum potential rate of photorespiration.

2.3.   THE PHYSICAL PROCESSES OF CARBON DIOXIDE EXCHANGE

The leaf is a heterogeneous assemblage of tissues, only parts of which are photosynthetically active, and assumptions need to be made about its structure and internal environment. If we assume that the photosynthetic sites, the chloroplasts, are uniformly distributed throughout the leaf tissue, and that within the leaf tissues there is a uniform light, carbon dioxide and oxygen environment, we can set up a two-compartment model of the leaf. Structurally we can treat the leaf as though it were an agglomeration of chlorenchyma tissues surrounded by a porous membrane. We can further assume that the activity of the photosynthetic sites within the chlorenchyma tissues is uniform and that inter-connected air spaces are dispersed throughout the chlorenchyma tissues. Water vapour, gaseous carbon dioxide and oxygen move between the bulk air surrounding the leaf and the internal air spaces by simple gaseous diffusion across the porous membrane. Carbon dioxide and oxygen then move between the internal air spaces and the photosynthetic sites.

The porous membrane, of course, represents the epidermal tissues of the leaf, the pores representing the stomates. The chlorenchyma tissue is the spongy

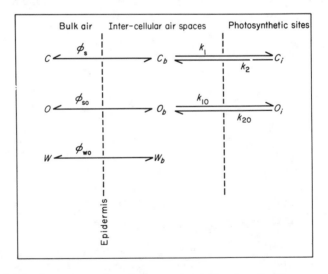

FIG. 2.2. Schematic representation of the movement of carbon dioxide, oxygen and water vapour between the bulk air and the photosynthetic sites within the leaf. C, O and W represent the concentrations of carbon dioxide, oxygen and water vapour. No suffix denotes the concentration in the bulk air, the suffixes b and i denote the concentrations in the intercellular air spaces and at the photosynthetic sites. The symbol ϕ denotes a diffusion constant and k denotes a transport constant.

and palisade mesophyll tissues of the leaves of C_3 plants and the bundle sheath cells of the leaves of C_4 plants. A simple scheme for the movement of water vapour, carbon dioxide and oxygen through the leaf, and based on these assumptions, is shown in Fig. 2.2. To simplify the mathematics we can also assume that the internal air spaces are saturated with water vapour at the leaf's temperature. At the steady state the net fluxes of water vapour, carbon dioxide and oxygen out of the leaf, and carbon dioxide and oxygen through the leaf, are constant, and the assumptions shown in Fig. 2.2 can be formally written as

$$\frac{dC}{dt} = P = P_G - (R_L + R_D) = \phi_s(C - C_b) = \phi_{so}(O_b - O) \quad (2.18)$$

$$\frac{dC_b}{dt} = 0 = \phi_s(C - C_b) - k_1 C_b + k_2 C_i \quad (2.19)$$

$$\frac{dC_i}{dt} = 0 = k_1 C_b - k_2 C_i - P \quad (2.20)$$

$$\frac{dO_b}{dt} = 0 = \phi_{so}(O_b - O) - k_{10}O_b + k_{20}O_i \quad (2.21)$$

$$\frac{dO_i}{dt} = 0 = P + k_{10}O_b - k_{20}O_i \quad (2.22)$$

and finally,

$$T = \phi_{wo}(W_b - W). \quad (2.23)$$

The symbols C, O and W represent concentrations of carbon dioxide, oxygen and water vapour in the bulk air surrounding the leaf, the subscripts b and i denote their concentrations in the internal air spaces and at the photosynthetic sites. T denotes the transpiration rate, the rate of loss of water vapour by the leaf. The symbols ϕ_s, ϕ_{so} and ϕ_{wo} are stomatal diffusion constants, k_1, k_2, k_{10} and k_{20} are transport constants.

Equations (2.18)–(2.22) can be solved, using quite simple algebra, to show that

$$C_i = (k_1 C_b - P)/k_2 = (k_1\phi_s C - (k_1 + \phi_s)P)/k_2\phi_s \quad (2.24)$$

and

$$O_i = (k_{10}O_b + P)/k_{20} = (k_{10}\phi_{so}O - (k_{10} - \phi_{so})P)/k_{20}\phi_{so}. \quad (2.25)$$

Provided that their diffusion pathways are the same, the diffusion constants for two different gases will be related by the ratio of their separate diffusion coefficients. For example, the stomatal diffusion constants for carbon dioxide, ϕ_s, and water vapour, ϕ_{wo}, are related by

$$\phi_{wo}/\phi_s \cong 1.7.$$

Provided that the leaf transpiration rate is known, the stomatal diffusion constant for carbon dioxide can be calculated from eqn (2.23) as

$$\phi_s = T/1.7(W_b - W) \tag{2.26}$$

where W is the ambient concentration of water vapour and W_b the saturating water vapour concentration at the leaf's temperature.

It is important to note that the rates of absorption and desorption of both carbon dioxide and oxygen in the cellular 'reaction solution' are implicit in the transport constants k_1, k_2, k_{10} and k_{20}. In effect, the solubility constants for carbon dioxide and oxygen are latent in the ratios k_1/k_2 and k_{10}/k_{20}. This may have some implications for our analysis of experimental data.

2.4. A MODEL FOR LEAF PHOTOSYNTHESIS

Now that we have developed simple mathematical descriptions of the biochemical and physical processes of photosynthesis we are in a position to construct a mathematical model of photosynthesis at the whole leaf level. The mathematics of eqns (2.10) and (2.11) can be simplified a little more by making two further assumptions. First, if we assume that the flux of oxygen through the leaf is small compared with the ambient oxygen concentration, and we presume that this assumption holds even when the ambient oxygen concentration is as low as 1% by volume or 13 g (O_2) m^{-3}, we can use eqn (2.25) to write

$$O_i = k_{10}O/k_{20} = K_oO. \tag{2.27}$$

Second, if we assume that δ and γ are sufficiently small so that all terms containing them can be ignored eqns (2.10)–(2.12) can be combined to give

$$P = \alpha_m Q(\tau C_i - \beta O_i)/(\alpha_m Q + \tau C_i) - R_D, \tag{2.28}$$

and using eqn (2.27),

$$P = \alpha_m Q(\tau C_i - \beta K_o O)/(\alpha_m Q + \tau C_i) - R_D. \tag{2.29}$$

Now eqn (2.24) gives the relationship

$$C_i = (k_1\phi_s C - (k_1 + \phi_s)P)/k_2\phi_s \tag{2.24}$$

and if we substitute for C_i in eqn (2.29), we obtain an expression of the form

$$P = [-b + \sqrt{(b^2 - 4ac)}]/2a \tag{2.30}$$

where the coefficients a, b and c are given by

$$a = -\tau(k_1 + \phi_s)/k_2\phi_s$$

$$b = \alpha_m Q + \tau k_1 C/k_2 - (R_D - \alpha_m Q)\tau(k_1 + \phi_s)/k_2\phi_s$$

$$c = R_D(\alpha_m Q + \tau k_1 C/k_2) - \alpha_m Q(\tau k_1 C/k_2 - \beta K_o O).$$

A simple check for eqn (2.30) is to put $Q = 0$, and then the coefficients become

$$a = -\tau(k_1 + \phi_s)/k_2 \phi_s$$
$$b = \tau k_1 C/k_2 - R_D \tau(k_1 + \phi_s)/k_2 \phi_s$$
$$c = R_D \tau k_1 C/k_2,$$

and on substitution into eqn (2.30) yield the identity

$$P = -R_D.$$

In the dark, of course, the rate of net photosynthesis or net carbon dioxide exchange by the leaf is equal to the dark respiration rate, the negative sign indicating the evolution of carbon dioxide.

We can derive several useful relationships from eqn (2.30). For example, when the net flux of carbon dioxide through the leaf is zero, that is when $P = 0$, eqn (2.30) reduces to

$$0 = R_D(\alpha_m Q + \tau k_1 C/k_2) - \alpha_m Q(\tau k_1 C/k_2 - \beta K_o O), \qquad (2.31)$$

and if we denote the ambient carbon dioxide concentration by Γ, rearrangement of eqn (2.31) leads to

$$\Gamma = k_2 \alpha_m Q(\beta K_o O + R_D)/k_1 \tau(\alpha_m Q - R_D). \qquad (2.32)$$

The physiological state when $P = 0$ is known as the leaf compensation point, and Γ is the compensation carbon dioxide concentration. As Q becomes large, so that $R_D \ll \alpha_m Q$, eqn (2.32) can be simplified to

$$\Gamma = k_2 (\beta K_o O + R_D)/k_1 \tau. \qquad (2.33)$$

At high light levels, or as Q becomes large and tends towards infinity, the rate of net photosynthesis becomes 'light saturated'. Equation (2.29) gives us

$$(P)_{Q \to \infty} = P_{MAX} = \tau C_i - \beta K_o O - R_D. \qquad (2.34)$$

Using eqn (2.24) to substitute for C_i in eqn (2.34) then allows us to write

$$P_{MAX} = (\tau k_1 C/k_2 - \beta K_o O - R_D)k_2 \phi_s/(k_2 \phi_s + \tau(k_1 + \phi_s)). \qquad (2.35)$$

We can now substitute for $(\beta K_o O + R_D)$ using eqn (2.33), and this leads to the expression

$$P_{MAX} = \phi_s k_1 \tau(C - \Gamma)/(\phi_s k_2 + \phi_s \tau + k_1 \tau). \qquad (2.36)$$

A third useful relationship can be obtained by the differentiation of P, in eqn (2.30), with respect to Q and taking the limit as Q tends to zero. This gives the initial slope of the light response curve for net photosynthesis, the leaf light utilization efficiency, α. We can write α as

$$\alpha = \alpha_m \left[1 - \beta K_o O k_1 \phi_s / \tau (k_1 \phi_s C + (k_1 + \phi_s) R_D) \right]. \tag{2.37}$$

It is clear from eqn (2.37) that α is a function of both the leaf's ambient carbon dioxide and oxygen concentrations. Leaf photosynthetic rates are usually expressed as the rates per unit leaf area. This basis for their calculation and expression comes about because the main natural environmental variable, the light received by the leaf, is commonly, and most easily, measured as the light flux density incident on the leaf's surface. If we denote the incident flux density of the photosynthetically active radiation (light in the 400-700 nm wavebands) by I, and assume that all of this light is absorbed by the leaf, we can write

$$Q = I/h \tag{2.38}$$

where h is the leaf's thickness. Note that we are assuming that the physical and reactive volumes of the leaf are the same, and also reinforcing our assumption in Section 2.3 that the photosynthetic sites are distributed uniformly throughout the volume of the leaf.

The rate of net photosynthesis per unit leaf area, F, can now be written as

$$F = Ph \tag{2.39}$$

and, for example, we can rewrite eqn (2.36) as

$$F_{MAX} = h \phi_s k_1 \tau (C - \Gamma)/(\phi_s k_2 + \phi_s \tau + k_1 \tau). \tag{2.40}$$

Indeed, we can obtain expressions for F from those for P by replacing the parameters ϕ_s, k_1, k_2, τ, β and R_D by the products $h\phi_s$, hk_1, hk_2, $h\tau$, $h\beta$ and hR_D $(= R_d)$, and the variable Q by I.

2.5. SOME PROPERTIES AND TESTS OF THE MODEL

When we examine the properties of the model developed in Section 2.4 it is useful to distinguish between C_3 and C_4 plant types (see Section 1.4). If we presume that the C_4 plants have biochemical processes that allow them to effectively 'concentrate' carbon dioxide at the photosynthetic sites, we can formalize this presumption in the model by assuming that for the C_4 plants $k_1 \gg k_2$. In contrast, for C_3 plants we could assume that movement of carbon dioxide between the inter-cellular air spaces and the photosynthetic sites follows the kinetics of a diffusive process, that is $k_1 = k_2 = \phi_m$, where we can call ϕ_m the mesophyll diffusion constant.

A. The C_3 Group of Plants

If we replace the transport constants k_1 and k_2 by the mesophyll diffusion constant ϕ_m, the coefficients a, b and c in eqn (2.30) become

$$a = -\tau(\phi_m + \phi_s)/\phi_m\phi_s$$

$$b = \alpha_m I/h + \tau C - (R_D - \alpha_m I/h)\tau(\phi_m + \phi_s)/\phi_m\phi_s$$

$$c = R_D(\alpha_m I/h + \tau C) - \alpha_m(I/h)(\tau C - \beta K_o O).$$

Equation (2.30) remains a quadratic in P. It describes a non-rectangular hyperbola for the response of the net photosynthetic rate to changes in ambient carbon dioxide and oxygen concentrations and curves similar, although not strictly hyperbolic, for the response to changing incident light flux density.

The behaviour of the model is best illustrated by 'fitting' it (see Section 1.2) to some experimental data. The response of the rate of net photosynthesis per unit leaf area, F, by leaves of sunflower (*Helianthus anuus*) to changes in the ambient oxygen and inter-cellular carbon dioxide concentrations at a high

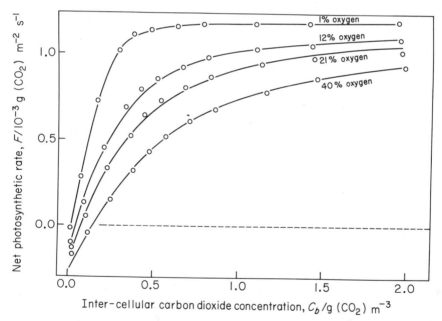

FIG. 2.3. Carbon dioxide response curves for the rate of net photosynthesis per unit leaf area of leaves of sunflower at four ambient oxygen concentrations and an incident light flux density of 120 W m^{-2} of photosynthetically active radiation. Solid curves result from fitting eqn (2.30) to the experimental data. Numerical values for the 'fitted' parameters α_m, $h\tau$, $h\beta K_o$ and $h\phi_m$ for these data are given in Table 2.1.

incident light flux density ($I = 120$ W m^{-2}) are shown in Fig. 2.3. The numerical values of the four coefficients α_m, $h\tau$, $h\beta K_o$ and $h\alpha_m$ giving rise to the 'fitted' response curves (the solid lines in Fig. 2.3) are given, together with their approximate 95% confidence intervals, in Table 2.1. The model provides a good visual description of these data.

TABLE 2.1. Numerical estimates of the photosynthesis parameters (α_m, $h\tau$, $h\beta K_o$ and $h\phi_m$), together with their approximate 95% confidence intervals, obtained by 'fitting' the leaf photosynthesis model to the data shown in Fig. 2.3.

α_m/g (CO_2) J^{-1}	10.0 (± 0.1) $\times$ 10^{-6}
$h\tau$/m s^{-1}	2.1 (± 0.2) $\times$ 10^{-3}
$h\beta K_o$/m s^{-1}	5.0 (± 0.7) $\times$ 10^{-7}
$h\phi_m$/m s^{-1}	4.8 (± 0.2) $\times$ 10^{-3}

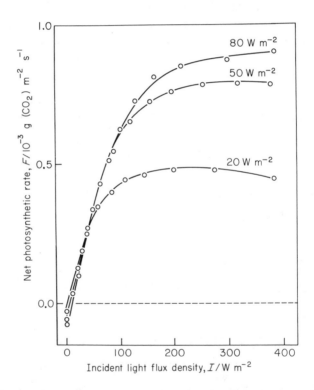

FIG. 2.4. Light response curves for the rate of net photosynthesis per unit leaf area of leaves of tomato at normal ambient carbon dioxide and oxygen concentrations. Leaves were from plants grown in contrasting light regimens (20, 50 and 80 W m^{-2} photosynthetically active radiation during a 16 h light period).

Similarly, the model can be 'fitted' to experimental data for the response of F to instantaneous changes in the light flux density incident on the leaf's surface. The 'fitted' curves, and experimental data for leaves of tomato (*Lycopersicon esculentum*) photosynthesizing at normal ambient carbon dioxide and oxygen concentrations (300 vpm or 0.59 g (CO_2) m^{-3} and 21% or 300 g (O_2) m^{-3}) are shown in Fig. 2.4. These responses were obtained for leaves from plants grown in three contrasting light regimes. The decline in F at high incident light flux densities, shown by the leaf from a plant grown in low light (20 W m^{-2}) is attributable to stomatal closure in this leaf at the higher light levels. Stomatal closure leads to a decrease in the value of ϕ_s. Estimates of $h\phi_s$ for all three leaves were obtained from leaf transpiration measurements (see eqn (2.26)), and the model was in effect 'fitted' using the inter-cellular carbon dioxide concentration, C_b, rather than the ambient concentration, as the independent variable.

Equation (2.32) defines the compensation value of the ambient carbon dioxide concentration, and if we relax the assumption that $k_1 = k_2 = \phi_m$, and instead write $k_1/k_2 = K_c$ (cf. eqn (2.27)), eqn (2.32) can be re-written as

$$\Gamma = \alpha_m I(\beta K_o O + R_D)/\tau K_c(\alpha_m I - hR_D). \tag{2.41}$$

Provided that $\alpha_m I \gg hR_D$, eqn (2.41) reduces to

$$\Gamma = (\beta K_o O + R_D)/\tau K_c \tag{2.42}$$

(note that we have replaced Q in eqn (2.32) by I/h, and hence the term hR_D appears in the denominator). Equation (2.42) now predicts that at high incident light flux densities the compensation concentrations of carbon dioxide and oxygen will be linearly related, and the slope of the relationship will be given by $\beta K_o/\tau K_c$. The slope describes the product of the ratio of the intrinsic photo-respiratory and photosynthetic activities of the leaf and what is, in effect, the ratio of the solubility constants for oxygen and carbon dioxide. The relationship between Γ and O, predicted by eqn (2.42), is in accord with experimental observation, and values of the slope of this relationship ($\beta K_o/\tau K_c$) obtained for

TABLE 2.2. Numerical estimates of the ratio $\beta K_o/\tau K_c$ for a number of C_3 plant species. Estimates were obtained from the slopes of the linear regressions of Γ on O in the temperature range 20–25°C.

Triticum aestivum	2.4×10^{-4}
Triticum vulgare	2.4×10^{-4}
Avena sativa	2.0×10^{-4}
Hordeum vulgare	2.4×10^{-4}
Phleum pratense	2.1×10^{-4}
Typha angustifolia	2.3×10^{-4}
Glycine max.	2.4×10^{-4}
Helianthus anuus	2.5×10^{-4}
Atriplex patula	2.3×10^{-4}

leaves from a number of C_3 plant species are given in Table 2.2. The numerical value of the slope is very sensitive to the leaf's temperature, but over a limited temperature range it is almost constant across the C_3 species shown in Table 2.2. The implications of this observation are discussed in Section 2.11.

B. The C_4 Group of Plants

If we assume that because of the PEP carboxylase reactions in these plants (see Section 1.4) the transport coefficient k_2 is far less than k_1, the coefficients a, b and c in eqn (2.30) can be re-written as

$$a = -\tau(k_1 + \phi_s)/k_2 \phi_s$$
$$b = \tau k_1 C/k_2 - (R_D - \alpha_m I/h)\tau(k_1 + \phi_s)/k_2 \phi_s$$
$$c = R_D \tau k_1 C/k_2 - \alpha_m I \tau k_1 C/h k_2.$$

Equation (2.30) remains quadratic in the rate of leaf net photosynthesis but the rate of net photosynthesis has become independent of the ambient oxygen concentration. Indeed, we can divide through by τ and $1/k_2$ to simplify the coefficients even further to

$$a = -(k_1 + \phi_s)/\phi_s$$
$$b = k_1 C - (R_D - \alpha_m I/h)(k_1 + \phi_s)/\phi_s$$
$$c = R_D k_1 C - \alpha_m I k_1 C/h.$$

The rate of leaf net photosynthesis at light saturation becomes, from eqn (2.36),

$$P_{MAX} = k_1 \phi_s (C - \Gamma)/(k_1 + \phi_s) \qquad (2.43)$$

but from eqn (2.33) we know that if $k_2 \ll k_1$, Γ will become very small so that

$$P_{MAX} = k_1 \phi_s C/(k_1 + \phi_s). \qquad (2.44)$$

Equation (2.44) can also be obtained from equation (2.30) by substitution of the coefficients a, b and c, shown above, and letting $I \to \infty$.

The prediction that the rate of net photosynthesis by leaves of C_4 plants will be independent of the ambient oxygen concentration is in accord with experimental observation. Indeed, this observation is one of the features of C_4 plant photosynthesis that distinguishes it from C_3 photosynthesis. Equation (2.32) defines the leaf carbon dioxide compensation concentration, Γ, according to

$$\Gamma = k_2 \alpha_m I(\beta K_o O + R_D)/\tau k_1(\alpha_m I - h R_D). \qquad (2.32)$$

If we assume that for leaves of C_3 plants $k_1 = k_2$, and that the parameters α_m, βK_o, R_D and τ have similar numerical values for leaves of C_3 and C_4 plants, we can write

$$(\Gamma)_{C_3}/(\Gamma)_{C_4} = (k_1/k_2)_{C_4}. \qquad (2.45)$$

(Note that even if $(k_1/k_2)_{C_3}$ is not unity, but is a 'solubility constant' for carbon dioxide in leaves of C_3 plants, the same 'solubility constant' will be implicit in the ratio $(k_1/k_2)_{C_4}$.) At normal ambient oxygen concentrations $(\Gamma)_{C_3}$ is about 0.1 g (CO_2) m^{-3} (about 50–60 vpm) and $(\Gamma)_{C_4}$ is about 0.05 g (CO_2) m^{-3} (about 2 or 3 vpm). We can use eqn (2.45) to calculate that the ratio $(k_1/k_2)_{C_4}$ is around twenty. That is, the transport constant for movement of carbon dioxide to the photosynthetic sites in leaves of C_4 plants is about twenty times greater than the constant for movement in the opposite direction. This reflects the assumption that the PEP carboxylase reactions constitute a carbon dioxide 'pump' in the leaves of C_4 plants.

2.6. EFFECTS OF LEAF TEMPERATURE

We can use eqn (2.37) to describe the initial slopes, α, of the light response curves for net photosynthesis by leaves of C_3 and C_4 plants. For a C_3 plant we have, if we assume that $k_1 = k_2 = \phi_m$,

$$(\alpha)_{C_3} = \alpha_m [1 - \beta K_o O \phi_m \phi_s / \tau(\phi_m \phi_s C + (\phi_m + \phi_s) R_D)] \qquad (2.46)$$

whereas for a C_4 plant

$$(\alpha)_{C_4} = \alpha_m. \qquad (2.47)$$

Equations (2.46) and (2.47) reflect the experimental observation that when the ambient oxygen concentration is reduced from 21% to 2% the initial slope of the light response curve for leaf net photosynthesis by C_3 plants increases, but for C_4 plants it does not change. It is also reported that the temperature dependence of the difference between the initial slopes in 21% and 2% oxygen is similar to the temperature dependence of the compensation carbon dioxide concentration in C_3 plants. In contrast, the initial slope of the light response curve for photosynthesis by leaves of C_4 plants is reported to be insensitive to temperature. If we assume that, at an ambient oxygen concentration of 2%, $(\alpha)_{C_3} \simeq \alpha_m$,

$$\frac{d}{dT} [(\alpha)_{C_3} - \alpha_m] = \frac{d}{dT} [\alpha_m \beta K_o O \phi_m \phi_s / \tau(\phi_m \phi_s C + (\phi_m + \phi_s) R_D)]. \qquad (2.48)$$

Equation (2.42) gives that at light saturation

$$\frac{d}{dT} (\Gamma)_{C_3} = \frac{d}{dT} [\beta K_o O + R_D)/\tau]. \qquad (2.49)$$

If R_D is small compared to both $\phi_m \phi_s C/(\phi_m + \phi_s)$ and $\beta K_o O$, and α_m is independant of temperature, eqns (2.48) and (2.49) lead to the identity

$$\frac{d}{dT}(\Gamma)_{C_3} = \frac{C}{\alpha_m} \frac{d}{dT}[(\alpha)_{C_3} - \alpha_m].$$ (2.50)

Equation (2.50) provides an explanation of the experimental observation that the effects of temperature on the initial slope of the light response curve for leaf net photosynthesis are similar to the effects on Γ for the C_3 plants.

It is useful if we now define an overall leaf carboxylation constant, Φ, such that for C_3 plants

$$\Phi = \phi_s \phi_m \tau / (\phi_s \phi_m + \phi_s \tau + \phi_m \tau).$$ (2.51)

Equation (2.40) can now be written as

$$F_{MAX} = h\Phi(C - \Gamma).$$ (2.52)

Let us now suppose that both Φ and Γ are temperature sensitive, and that their sensitivities can be described by Arrhenius rate equations. The two assumptions enable us to write

$$\Phi = \Phi_0 \exp(-E_\Phi / RT)$$ (2.53)

and

$$\Gamma = \Gamma_0 \exp(-E_\Gamma / RT)$$ (2.54)

where T is the leaf temperature in degrees Kelvin, R is the gas constant, E_Φ and E_Γ are 'apparent activation energies' for Φ and Γ, with Φ_0 and Γ_0 their extrapolated values at 0 K. Substitution for Γ and Φ in eqn (2.52) gives

$$F_{MAX} = h\Phi_0 \exp(-E_\Phi / RT) [C - \Gamma_0 \exp(-E_\Gamma / RT)].$$ (2.55)

If E_Φ and E_Γ are given values of 10 kcal mol^{-1} and 20 kcal mol^{-1} respectively, and Φ_0 and Γ_0 chosen so that $h\Phi$ and Γ have values of 1.5×10^{-3} m s^{-1} and 1.00×10^{-4} kg (CO_2) m^{-3} at 20°C (293.2 K), and we assume a constant ambient carbon dioxide concentration of 0.6×10^{-3} kg (CO_2) m^{-3} (~300 vpm), eqn (2.48) can be used to simulate the temperature response of F_{MAX} for a C_3 plant. The results of this simulation are shown in Fig. 2.5. The simulation suggests that a maximum value of F_{MAX} for a C_3 plant could arise when the respiratory production of carbon dioxide (both in R_L and R_D and reflected in eqn (2.52) in the value of Γ) becomes equal to its gross rate of consumption in photosynthesis. The hypothesis can be usefully explored a little more. At low temperatures, when $\Gamma \ll C$, the temperature response of F_{MAX} reflects that of Φ. Mathematically we can write that

$$F_{MAX} \simeq h\Phi C = h\Phi_0 C \exp(-E_\Phi / RT).$$ (2.56)

At higher temperatures, we could, by extrapolation of eqn (2.56), followed by

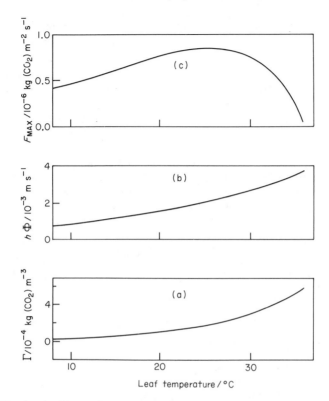

FIG. 2.5. Simulated effects of leaf temperature on (a) compensation carbon dioxide concentration (Γ), (b) overall leaf conductance ($h\Phi$) and (c) the rate of net photosynthesis per unit leaf area at light saturation (F_{MAX}). Values of –10 kcal mol^{-1} and –20 kcal mol^{-1} are assumed for E_Φ and E_Γ.

subtraction of the 'observed' values given by eqn (2.55), obtain the expression

$$(F_{MAX})_{extrapolated} - (F_{MAX})_{observed} = h\Phi_0\Gamma_0 \exp[-(E_\Phi + E_\Gamma)/RT]. \quad (2.57)$$

When the leaf temperature is sufficiently high that $C \ll \Gamma$, we can obtain directly from eqn (2.55) the relationship

$$F_{MAX} = -h\Phi_0\Gamma_0 \exp[-(E_\Phi + E_\Gamma)/RT]. \quad (2.58)$$

It follows from eqns (2.56)–(2.58) that provided the numerical values of E_Φ and E_Γ remain constant over the entire temperature range, their values should be obtainable from an analysis of the different parts of the experimental response of F_{MAX} to temperature. Equation (2.58) will also hold if the ambient carbon dioxide concentration is maintained at zero, that is if we study the evolution of carbon dioxide into carbon dioxide free air.

The approach has been used to examine experimental data on the temperature responses of steady-state, light saturated, photosynthesis by C_3 grasses. A re-analysis of some published data is shown in Table 2.3. The physical interpretation of the numerical values of E_Φ, E_Γ and $(E_\Phi + E_\Gamma)$ cannot be meaningfully attempted, but the data do lend some support to the hypothesis that the maximum value of F_{MAX} observed experimentally is caused by the respiratory rate of the leaves of C_3 plants becoming comparable with their gross photosynthetic rates.

TABLE 2.3. (a) Mean values of the parameters E_Φ and E_Γ, estimated from the temperature response curves of F_{MAX} using eqns (2.56)–(2.58), for *Lolium perenne* and *Lolium multiflorum*. (b) Mean values of $(E_\Phi + E_\Gamma)$ for six plant species estimated, using eqn (2.58), from the evolution of carbon dioxide into carbon dioxide free air in the light.

(a)

	E_Φ/kcal mol^{-1}	E_Γ/kcal mol^{-1}
Lolium perenne	18.1	13.1
Lolium multiflorum	13.3	15.3

(b)

	$(E_\Phi + E_\Gamma)$/kcal mol^{-1}
Lolium perenne	15.2
Lolium multiflorum	7.2
Atriplex nummularia	10.8
Atriplex hastata	11.0
Glycine max.	12.5
Sugar beet	10.0

2.7. SOME EFFECTS OF LEAF WATER DEFICITS

As the soil which surrounds a plant's roots dries, and the soil water deficit increases, the plant usually begins to experience some degree of water stress. Most plants can ameliorate both the rate of onset and extent of the stress by reducing the rates at which their leaves lose water. There is considerable evidence to suggest that the rate of leaf water loss, the leaf transpiration rate, of these plants is reduced by stomatal closure. In these plants the stomatal conductance ($h\phi_s$) appears to decrease as the leaf water potential (ψ_ϱ) becomes more negative. This effect can be modelled mathematically if it is assumed that the stomatal conductance is linearly related to the water content of the stomates' guard cells, and that changes in the water content of the guard cells are due to metabolically-controlled changes in their osmotic pressure. The model gives rise to three linear relationships between $h\phi_s$ and ψ_ϱ over the physiologically observable range of leaf water potentials. These relationships are

$$h\phi_s = h(\phi_s)_0, \qquad \psi_\varrho > \psi_0 \qquad (2.59)$$

$$h\phi_s = h(\phi_s)_0(2\psi_{1/2} - \psi_0 - \psi_\varrho)/2(\psi_{1/2} - \psi_0),$$
$$\psi_0 \gg \psi_\varrho > (2\psi_{1/2} - \psi_0) \qquad (2.60)$$

$$h\phi_s = 0, \qquad \psi_\varrho < (2\psi_{1/2} - \psi_0), \qquad (2.61)$$

where ψ_0 is the value of ψ_ϱ above which $h\phi_s$ shows some maximum value $h(\phi_s)_0$ and no longer responds to changes in ψ_ϱ, and $\psi_{1/2}$ is that value of ψ_ϱ at which $h\phi_s = \frac{1}{2}h(\phi_s)_0$.

These equations, and more particularly eqn (2.60), have been used to analyse a variety of experimental data, but because of their discontinuous behaviour they have some drawbacks and it is necessary to use one of two approaches in their application. Either judgements can be made about the numerical values of the quantities ψ_0 and $(2\psi_{1/2} - \psi_0)$, and hence which data sets to exclude when using eqn (2.60) to regress $h\phi_s$ on ψ_ϱ, or the j data pairs ($h\phi_s$ and ψ_ϱ) can be

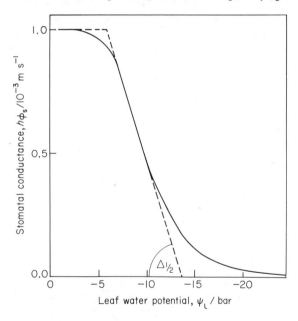

FIG. 2.6. A comparison of the behaviours of eqns (2.59)–(2.61), broken lines, and eqn (2.62), solid line. In both simulations $h(\phi_s)_0$ and $\psi_{1/2}$ were set at 1×10^{-2} m s^{-1} and –10 bars respectively. With eqns (2.59)–(2.61) ψ_0 was set at –6 bars, and for eqn (2.62) n was set at 5. The values are not entirely arbitrary, the sensitivity of $h\phi_s(\Delta_{1/2})$ is the same in both simulations.

arranged in ascending order of the independent variable ψ_ϱ, and linear regressions 'fitted' to the first i pairs and the next k pairs and remaining $j - (k + i)$ data pairs, adjusting the numbers of the i and k pairs until the combined residual sum of squares of the three regressions is a minimum.

An alternative empirical function, which is continuous and has similar properties to eqns (2.59)-(2.61), can be written as

$$h\phi_s = h(\phi_s)_0/(1 + (\psi_\varrho/\psi_{1/2})^n), \qquad \psi_\varrho < 0, \qquad (2.62)$$

where $h(\phi_s)_0$ and $\psi_{1/2}$ are as defined above and n is a constant. Equations (2.59)-(2.61) and eqn (2.62) are compared in Fig. 2.6. It is common to talk of the 'sensitivity' of stomatal conductance to changes in the leaf water potential. The 'sensitivity' can be defined as the slope of eqn (2.62) at $\psi_{1/2}$, and denoted by $\Delta_{1/2}$, such that

$$\Delta_{1/2} = -nh(\phi_s)_0 /4\psi_{1/2}. \qquad (2.63)$$

The analytical application of eqns (2.62) and (2.63) is demonstrated in Fig. 2.7 and Table 2.4. Plants of the forage legume *Macroptilium atropurpureum* (cv. Siratro) were grown in pots in a controlled environment room and subjected to a series of seven successive droughting cycles. This was accomplished simply by watering the plants and then allowing the pots to 'dry out'. One set of data for the dependence of $h\phi_s$ on ψ_ϱ is shown in Fig. 2.7 along with the 'fitted'

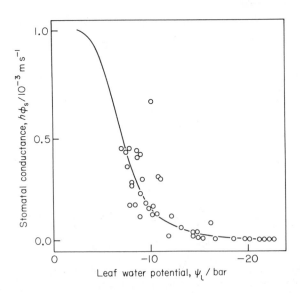

FIG. 2.7. The results of 'fitting' eqn (2.62) to data for leaves of the legume *Macroptilium atropurpureum*. A value for $h(\phi_s)_0$ of 1×10^{-2} m s^{-1} was assumed when fitting the model.

TABLE 2.4. Estimated numerical values of the parameters n and $\psi_{1/2}$ obtained by 'fitting' eqn (2.62) to experimental data for leaves of the legume *Macroptilium atropurpureum*. Leaves were from plants subjected to a succession of droughting cycles. For all data sets a value of $h(\phi_s)_0$ of 0.01 m s^{-1} was assumed. The stomatal 'sensitivity', $\Delta_{1/2}$, calculated for each pair of values of n and $\psi_{1/2}$ is also shown.

Number of droughting cycles experienced by plant	n	$\psi_{1/2}$/bar	$\Delta_{1/2}$/m s^{-1} bar^{-1}
1	5.3 (1.7)	−7.4 (1.2)	0.19
3	4.6 (0.8)	−6.8 (0.8)	0.18
5	5.3 (0.7)	−7.3 (0.7)	0.19
7	5.6 (1.0)	−7.0 (0.9)	0.21

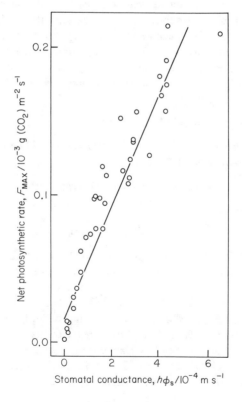

FIG. 2.8. The regression of F_{MAX} on $h\phi_s$ for leaves of the forage legume *Macroptilium atropurpureum* towards the end of a droughting cycle.

TABLE 2.5. Numerical values of the coefficients m and c obtained for the linear regression of F_{MAX} on $h\phi_s$ ($F_{MAX}= m(h\phi_s) + c$) for leaves of the legume *Macroptilium atropurpureum*. The leaves were from plants subjected to a succession of droughting cycles. (R^2 denotes the percent variance accounted for by the regression.)

Number of droughting cycles experienced by plant	$m/10^3$ μmol m^{-3}	$c/\mu mol$ m^{-2} s^{-1}	R^2
1	1.95 (0.27)	1.75 (0.70)	88
3	2.98 (0.32)	1.29 (0.82)	91
5	3.73 (0.44)	1.76 (1.09)	89
7	3.27 (0.32)	1.19 (0.76)	92

relationship predicted by eqn (2.62). Fitted values of the parameters n and $\psi_{1/2}$ obtained by analysis of data derived from each of the successive droughting cycles, and values of the 'responsiveness' of the stomates during each cycle, are shown in Table 2.4. The legume does not appear to adapt to a series of droughting cycles, and this may be of some agronomic interest.

Most 'field data' is of the sort shown in Fig. 2.7, values of $h\phi_s$ at the higher leaf water potentials not being commonly observed. If $h\phi_s \ll h(\phi_s)_0$, eqn (2.62) reduces to the form

$$\ln(h\phi_s) = n \ln(|\psi_\varrho|) + n \ln(|\psi_{1/2}|) - \ln(h(\phi_s)_0). \qquad (2.64)$$

Equation (2.64) provides a robust, linear analysis for 'field data'.

Now we can re-write eqn (2.40) as

$$F_{MAX} = h\phi_s\Phi^*(C - \Gamma)/(\phi_s + \Phi^*) \qquad (2.65)$$

where Φ^* is defined by

$$\Phi^* = k_1\tau/(k_2 + \tau). \qquad (2.66)$$

(Note that $\phi_s\Phi^*/(\phi_s + \Phi^*)$ is equivalent to Φ in eqn (2.51).) Clearly if ϕ_s decreases sufficiently so that $\phi_s \ll \Phi$, eqn (2.65) reduces to

$$F_{MAX} = h\phi_s(C - \Gamma), \qquad (2.67)$$

which predicts that F_{MAX} should be linearly dependant on $h\phi_s$. This prediction is borne out by the data from *M. atropurpureum* plants subjected to soil water deficits, and is illustrated in Fig. 2.8. Data on F_{MAX} and $h\phi_s$ obtained during the later stages of each of the successive droughting cycles were subjected to linear regression analysis. The results are shown in Table 2.5.

2.8. AN EXTENSION OF THE MODEL

A mathematical model of the instantaneous response of the rate of leaf net photosynthesis to changes in the leaf's environment has been described in the previous sections. It is now of interest to see how this model can be used to analyse and order some of the genetic and adaptive variation in leaf photosynthetic activity. Since both genetic and adaptive variability can be shown to exist in leaf structure it is necessary to be aware how this source of variability might affect the leaf's photosynthetic behaviour. In this section, therefore, the model for leaf photosynthesis is extended to incorporate some of the more readily distinguishable and measurable characters of leaf structure and anatomy.

It has been demonstrated that for a leaf of a C_3 plant the rate of light-saturated net photosynthesis per unit of leaf area, F_{MAX}, can be written as

$$F_{MAX} = h\phi_s\phi_m\tau(C - \Gamma)/(\phi_s\phi_m + \phi_s\tau + \phi_m\tau) \tag{2.68}$$

where h is the leaf thickness (see eqns (2.51) and (2.52)). The symbols ϕ_s and ϕ_m denote stomatal and mesophyll diffusion constants, and are likely to be functions of leaf structure and anatomy. It is useful to remind ourselves of Fick's Law of gaseous diffusion. Fick's Law can be written down as

$$\Delta W/\Delta t = -DA^* \Delta M/\Delta X \tag{2.69}$$

where $\Delta W/\Delta t$ is the flux of substance M across a boundary with cross-sectional area A^*, $\Delta M/\Delta X$ is the concentration gradient of substance M across the boundary of thickness ΔX and D is a diffusion coefficient. If we assume a homogeneous environment in M on either side of the boundary we can replace $D/\Delta X$ by D^*, an effective diffusion coefficient which incorporates the finite thickness, ΔX, of the boundary. Now eqn (2.18) allows us to write that at light saturation

$$F_{MAX} = h\phi_s(C - C_b) \tag{2.70}$$

where C is the ambient, and C_b the intercellular, carbon dioxide concentration. By analogy with eqn (2.69) we can write

$$AF_{MAX} = D_s^*A^*(C - C_b) \tag{2.71}$$

where A is the leaf area (and AF_{MAX} is the flux of carbon dioxide out of the leaf) and D_s^* an effective stomatal diffusion coefficient. We can now add, using eqns (2.70) and (2.71), that

$$\phi_s = D_s^*A^*/hA, \tag{2.72}$$

where A^* is the actual cross-sectional area of the boundary across which the carbon dioxide diffuses. If there are p stomates per unit of leaf area, and each

TABLE 2.6. Some stomatal characteristics of contrasting genotypes of *Lolium perenne*. The variety C_1 has shallow epidermal ridging whilst the variety C_2 has deep ridging. The stomatal conductance measurements, $h\phi_s$, are derived from measurements of the leaf diffusion resistance. The stomatal frequency, p, and mean stomatal pore length, L, were measured directly.

	$h\phi_s/10^{-2}$ m s^{-1}	$p/10^6$ m^{-2}	$L/10^{-6}$ m
C_1			
Adaxial surface	0.22	106	24.6
Abaxial surface	0.01	18	20.3
C_2			
Adaxial surface	0.29	121	22.4
Abaxial surface	0.06	41	21.2

has a mean cross-sectional area a, we can write

$$A^* = paA \tag{2.73}$$

so that eqn (2.73) becomes

$$\phi_s = D_s^* pa/h. \tag{2.74}$$

The stomatal resistance, r_s, is the reciprocal of the product $\phi_s h$. We can write, using eqn (2.74), that

$$1/r_s = D_s^* pa, \tag{2.75}$$

which predicts a linear relationship between $(1/r_s)$ and both p and a. The reciprocals of the stomatal resistances for the adaxial and abaxial surfaces of leaves of two constrasting *Lolium* genotypes provide some support for eqn (2.75). These data, shown in Table 2.6, suggest that large differences between

TABLE 2.7. Environmentally induced differences in the stomatal conductance, $h\phi_s$ and stomatal frequences, p, and estimated thickness, h, of leaves of tomato (*Lycopersicon esculentum*). Plants were grown at three contrasting light levels during a 16 h light period. Derived values of the stomatal diffusion constant, ϕ_s, are also shown.

Growth light level/W m^{-2}	20	50	100
$h\phi_s/10^{-3}$ m s^{-1}	3.1	5.2	5.7
$p/10^6$ m^{-2}			
Adaxial	42	117	152
Abaxial	173	262	269
$h/10^{-4}$ m	1.5	2.0	2.8
ϕ_s/s^{-1}	21	26	20

the reciprocals of the stomatal resistances for the adaxial and abaxial leaf surfaces are associated with large differences in the stomatal densities, p. The mean length of the stomatal pore, which provides some indication of the mean stomatal cross-sectional area, a, does not appear to alter markedly. Environmentally induced, adaptive, differences in the reciprocal of the stomatal resistance of tomato leaves also appear to be associated with differences in stomatal densities. These data, shown in Table 2.7 and obtained for leaves from plants grown at different light levels, show a 50% change in the abaxial stomatal density compared with a four fold change in the adaxial density. The adaptive response of the stomatal conductance (the reciprocal of the stomatal resistance) is also associated with leaf thickening. As a result the stomatal diffusion constant, ϕ_s, does not appear to change with growth light level, and the time for movement of unit amount of carbon dioxide from the air to the inter-cellular air spaces remains constant despite the considerable differences in leaf thickness.

We can adopt a similar approach to study the effects of structural changes on the mesophyll diffusion constant, ϕ_m. We need first to make some assumptions about the geometry of the mesophyll cells. If we assume that they are spherical, with mean radius r, and that there are m cells beneath unit leaf area, the effective area of the boundary surface across which carbon dioxide diffuses when moving from the inter-cellular air spaces to the photosynthetic sites is

$$A^* = Am(4\pi r^2). \tag{2.76}$$

Following eqn (2.74) we can now write

$$\phi_m = D_m^*(4\pi r^2)/h, \tag{2.77}$$

where D_m^* is an effective mesophyll diffusion coefficient. If we denote the mean mesophyll cell size, that is the mean cell cross-sectional area, by σ, then

$$\sigma = \pi r^2 \tag{2.78}$$

and

$$\phi_m = D_m^* 4m\sigma/h. \tag{2.79}$$

Both genetic and adaptive variation in *Lolium* can give rise to an inverse relationship between m and σ, and this is demonstrated in Fig. 2.9. The product $m\sigma$ also appears to be related to the thickness of the leaf's mesophyll tissue, which is presumably related to the leaf's overall thickness. This is also shown in Fig. 2.9. If D_m^* remains constant in all these leaves, Fig. 2.9 implies that $m\sigma$ will also remain constant. However, the relationship implied by Fig. 2.9 needs to be treated with some caution. We could write eqn (2.76) as

$$A^* = Vq(4\pi r^2) = Ahq(4\pi r^2), \tag{2.80}$$

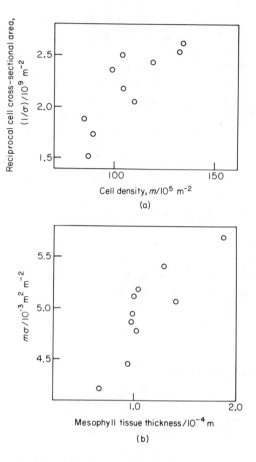

FIG. 2.9. The observed relationships between (a) the reciprocal of mean mesophyll cross-sectional area (σ) and the number of mesophyll cells per unit leaf area (m) and (b) the product $m\sigma$ and mean mesophyll tissue thickness arising from genetic and environmental variability in seedlings of *Lolium perenne*.

where V is the volume of the leaf and q the number of mesophyll cells per unit leaf volume. Now, if mesophyll cells are spherical we might expect that the larger their radius, the fewer will pack into unit leaf volume. We could then write

$$A^* = bV/r = bAh/r, \qquad (2.81)$$

where b is a proportionality constant which depends on the packing arrangement of the cells. Combining eqns (2.76) and (2.81) then gives

$$Am(4\pi r^2) = bAh/r$$

and thence

$$m = bh/4\sigma r. \tag{2.82}$$

It should now be apparent that in plotting the product $m\sigma$ against h (eqn (2.79)) there is the danger of an artificial correlation, and the plot is really equivalent to one of h/r against h.

If we use eqn (2.82) to substitute for m in eqn (2.79) we have

$$\phi_m = D_m^* \, b/r, \tag{2.83}$$

which provides an inverse relationship between ϕ_m and the mean mesophyll cell radius. If the leaf thickness is proportional to the mean mesophyll cell radius, that is

$$h = kr \tag{2.84}$$

where k is a constant, then eqn (2.82) can be re-written as

$$m = bk/4\sigma, \tag{2.85}$$

which predicts the relationship shown in Fig. 2.9, and eqn (2.79) becomes

$$\phi_m = D_m^* bk/h. \tag{2.86}$$

If we now use eqns (2.74) and (2.86) to substitute for ϕ_s and ϕ_m in eqn (2.68) we have

$$F_{MAX} = h\tau(C - \Gamma)/(1 + Ah\tau) \tag{2.87}$$

where

$$A = (D_s^* pa + D_m^* kb)/D_s^* pa D_m^* kb. \tag{2.88}$$

Equation (2.87) predicts that F_{MAX} will be hyperbolically dependent on both leaf thickness, h, and the carboxylation constant τ.

A note of caution needs to be raised here. We are implicitly assuming that the overall physical volume of the leaf is proportional to the volume of the leaf's photosynthetic apparatus. This may not always be true, and some other measure, such as specific leaf area (leaf area per unit leaf weight), may provide a better indicator of the leaf's 'metabolic' volume.

2.9. LEAF ADAPTATION TO LIGHT AND TEMPERATURE

The ways in which a leaf adapts to changes in the plant's growth environment can be expressed at several different levels of the leaf's organization. For example, whereas the environment experienced by the leaf during growth and development might affect both the size and specific activity of its

TABLE 2.8. Effects of different light regimens during growth and development on the rates of light saturated photosynthesis per unit leaf area, F_{MAX}, leaf thickness, h (or mesophyll tissue thickness, h^*) and on the rates of light saturated photosynthesis per unit leaf volume, P_{MAX} (or per unit mesophyll tissue volume, P^*_{MAX}).

	$h/10^{-3}$ m	$h^*/10^{-3}$ m	$F_{MAX}/10^{-3}$ g m^{-2} s^{-1}	$P_{MAX}/$g m^{-3} s^{-1}	$P^*_{MAX}/$g m^{-3} s^{-1}
Plectranthus parviflorus					
1 (E m^{-2})	0.28	—	0.30	1.1	—
7	0.51	—	0.56	1.1	—
23	0.83	—	0.70	0.9	—
Lycopersicon esculentum					
1 (MJ m^{-2})	0.15	—	0.48	3.1	—
3	0.20	—	0.79	4.0	—
5	0.28	—	0.89	3.2	—
Lolium perenne					
3 (MJ m^{-2})	0.21	—	0.38	1.9	3.6
6	0.20	—	0.42	2.1	3.8
14	0.19	—	0.47	2.5	4.5
Abutilon theophrasti					
5 (E m^{-2})	—	0.07	0.36	—	5.1
17	—	0.09	0.55	—	6.1
41	—	0.12	0.66	—	6.1
Gossypium hirsutum					
5 (E m^{-2})	—	0.11	0.24	—	2.2
17	—	0.19	0.43	—	2.3
41	—	0.28	0.79	—	2.8

photosynthetic apparatus, when growth has stopped the environment is only likely to affect the specific activity of the apparatus. Essentially, the size of the leaf's photosynthetic apparatus is primarily determined during growth and development of the remainder of the leaf. The analysis developed in the preceding section provides a basis for the examination of leaf adaptation during growth. It is important to re-emphasize the caution given in the last paragraph of that section. It is implicitly assumed that the physical volume of the leaf is proportional to its 'metabolic' volume and, if there are gross changes in leaf structure, this may not be true. Specific leaf area† (leaf area per unit of leaf weight) might be a better indicator, than overall leaf thickness, of the 'metabolic' volume beneath unit leaf area. However, since some plants can accumulate large quantities of starch in granules within the chloroplast (up to 15% of the leaf's dry weight), even specific leaf area might not be entirely adequate.

The light and temperature regimens of the plant during leaf growth and development both affect leaf thickness, and presumably the size of the photosynthetic apparatus beneath unit leaf area. The effects of different light levels on the rate of light saturated photosynthesis per unit leaf area, F_{MAX}, per unit leaf volume, P_{MAX}, and leaf thickness, h, in five species are shown in Table 2.8. Generally both h and F_{MAX} increase with increasing light integrals during leaf growth and development. In contrast, P_{MAX} remains fairly constant across the growth light levels. This observation implies that the size of the photosynthetic apparatus is affected by the growth light regimen, but not the leaf's specific photosynthetic activity. In the C_3 grass *Lolium perenne,* however, h scarcely alters, and there is some indication that P_{MAX} increases. In *Fragaria virginiana,* whilst F_{MAX} increases with increasing growth light levels, h remains fairly constant. However, the specific leaf area, s_A, decreases, and the rate of light saturated photosynthesis per unit leaf weight ($s_A \times F_{MAX}$) does not change significantly with the growth light level. These data are shown in Table 2.9. It may be that the specific leaf area is a better indication of the size of the 'metabolic' volume beneath unit leaf area in these leaves and that the specific leaf

---

† 'Specific activity' conventionally means 'activity per unit weight of system'. Specific leaf area is accordingly taken throughout as the leaf area per unit leaf dry weight, that is

$$s_A = A_L/W_L$$

where A_L is the leaf area and W_L is the leaf weight. The term specific leaf weight, often used to describe the reciprocal of s_A, is not used here. The prefix specific is used consistently to describe 'amount' or 'activity' per unit dry weight. If we assume that the density of plant dry matter, ρ, is constant the leaf 'metabolic' volume can be written as

$$V_L = W_L/\rho$$

and hence

$$s_A = 1/h'\rho,$$

where h' is the leaf's 'metabolic' thickness.

TABLE 2.9. Effects of different light regimens during growth and development on the rates of light saturated photosynthesis per unit leaf area, F_{MAX}, and specific leaf area (s_A) of leaves of *Fragaria virginiana*. Values of F_{MAX}, s_A and the product $F_{MAX}s_A$ followed by different letters are statistically significantly different.

S/E m^{-2}	$F_{MAX}/10^{-3}$ g m^{-2} s^{-1}	$s_A/$m^2 g^{-1}	$F_{MAX}s_A/10^{-3}$ g g^{-1} s^{-1}
6.5	1.0 (a)	0.021 (a)	0.022 (a)
9.9	1.2 (b)	0.020 (a)	0.025 (a)
10.1	0.9 (a)	0.023 (a)	0.022 (a)
16.3	1.3 (b)	0.017 (b)	0.020 (a)
20.0	1.4 (b)	0.015 (c)	0.020 (a)

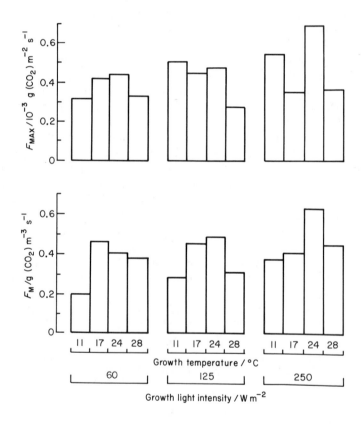

FIG. 2.10. The effects of different growth light and temperature regimens on the rates of light saturated photosynthesis per unit leaf area, F_{MAX}, and per unit volume of mesophyll tissue, F_M, by leaves of seedlings of *Lolium perenne*.

TABLE 2.10. Effects of different temperature regimens during growth and development on the rates of light saturated photosynthesis per unit leaf area, F_{MAX}, mean leaf thickness, h, and the rates of light saturated photosynthesis per unit leaf volume, P_{MAX}.

$T/^\circ C$	$h/10^{-3}\,m$	$F_{MAX}/10^{-3}\,g\,m^{-2}\,s^{-1}$	$P_{MAX}/g\,m^{-3}\,s^{-1}$
Plectranthus parviflorus			
6.5	0.48	0.34	0.71
17.5	0.51	0.56	1.10
28.5	0.54	0.72	1.33
Lolium perenne			
11.0	0.25	0.45	1.80
17.0	0.19	0.40	2.00
24.0	0.20	0.53	2.70
28.0	0.17	0.32	2.00

photosynthetic activity is not affected by the growth light environment.

With leaves of *Plectranthus parviflorus* increased growth temperature environments give rise to thicker leaves and increased rates of light saturated photosynthesis on both an area and volume basis. In contrast, higher growth temperatures lead to thinner leaves of *L. perenne*. In these plants F_{MAX} is higher for leaves on plants grown at 11°C compared with plants grown at 17° and 28°C. However, it can be seen from Table 2.10 that P_{MAX} for leaves of *L. perenne* increase with increasing growth temperature up to 24° C. The apparent stimulation of F_{MAX} at low growth temperatures is also evident in leaves of *D. glomerata* and *Chrysanthemum morifolium* (see Table 3.3). This stimulation could be attributed to a simple increase in the size of the photosynthetic apparatus beneath unit leaf area with decreasing growth temperatures, although the specific activity of that apparatus may decline.

In *L. perenne* there is an interaction between the growth light and temperature regimens. The rates of light saturated photosynthesis per unit leaf area and per unit volume of leaf mesophyll tissue for *L. perenne* plants grown in contrasting light and temperature regimens are shown in Fig. 2.10. At the two higher growth light levels there is a marked increase in F_{MAX} at the low growth temperature (11°C). However, the rate of photosynthesis per unit volume of mesophyll tissue, P^*_{MAX}, shows no such increase. The apparent stimulation in F_{MAX} can be attributed to leaf thickening.

2.10. OTHER EFFECTS

The light and temperature regimens during plant growth are not the only effectors of the adaptive responses of the leaf photosynthetic apparatus. For example, some plants adapt to sustained soil water deficits. They do so by modifying the response of their leaf conductance to decreasing leaf water potential, and this form of adaptation could be studied using the analysis suggested in Section 2.7. One source of considerable environmental variation is the fertility of the rooting medium, i.e. soil fertility. One of the most readily studied elements, acquired by the plant's roots from the soil, is nitrogen. Since proteins contain nitrogen, and are the catalytic machinery of the living cell,

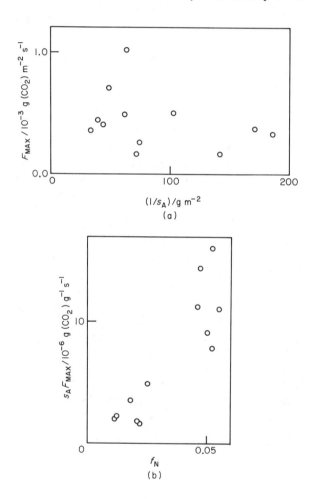

FIG. 2.11. The relationships between (a) the rate of light saturated photosynthesis, F_{MAX}, and the reciprocal of the specific leaf area and (b) the product $F_{MAX} \cdot s_A$ and leaf nitrogen content, f_N, for leaves from *Eucalyptus* species originating from different growth habitats.

TABLE 2.11. The effects of different levels of nitrogen application on the total dry weight, W, the proportion of the dry weight in shoots, η_S, specific leaf area, s_A, rate of light saturated photosynthesis per unit leaf area, F_{MAX} and nitrogen content of dry matter in 6-week-old seedlings of *Lolium multiflorum*.

Nominal application level of Nitrogen	W/g	η_S	$F_{MAX}/10^{-3}$ g m^{-2} s^{-1}	s_A/m^2 g^{-1}	$f_N/10^{-2}$	$s_A F_{MAX}/10^{-6}$ g g^{-1} s^{-1}
8 ppm	1.2	0.37	0.22	0.019	0.5	4.2
23 ppm	2.9	0.46	0.49	0.014	0.9	7.0
68 ppm	9.1	0.53	0.55	0.013	1.2	7.1
203 ppm	12.7	0.63	0.61	0.021	1.8	12.7

nitrogen deficiency in the rooting medium might be expected to have a dramatic effect on plant growth.

If seedlings of the C_3 grass *Lolium multiflorum* are grown in washed sand, and nutrient solution, containing different amounts of ammonium nitrate, is applied at 2–3 day intervals, after 6 weeks there are large differences in both the seedling dry weights and the proportions of the dry weight contained in the shoots. Data for seedlings of *L. multiflorum* grown in this way are shown in Table 2.11. The rates of light saturated photosynthesis per unit leaf area and specific leaf areas of the youngest fully expanded leaf of seedlings from each treatment are also shown in the table. Both F_{MAX} and the product $s_A F_{MAX}$ increase with increasing levels of nitrogen application. The data suggest that nitrogen availability affects the specific photosynthetic activity of the leaves, and $s_A F_{MAX}$ then appears to be correlated with the nitrogen content of the dry matter. At low levels of available nitrogen the plants have grown less well, they have partitioned more of their dry matter to the roots, presumable to 'scavenge' for more nitrogen, and their specific photosynthetic activity is reduced.

We observe somewhat similar behaviour in *Eucalyptus* species originating

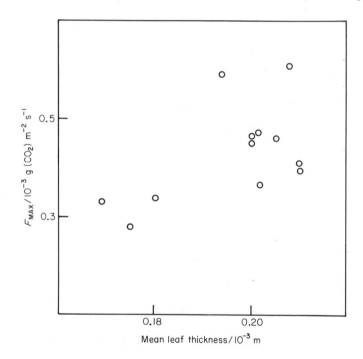

FIG. 2.12. The relationship between the rate of light-saturated photosynthesis, F_{MAX}, and mean leaf thickness for contrasting genotypes of *Lolium perenne*.

from Australian habitats differing in their moisture regimes. Although grown in this experiment under the same nutrient regimes the rates of light saturated photosynthesis per unit leaf area, F_{MAX}, of the different species do not appear to be correlated with differences in their specific leaf areas (see Fig. 2.11). However, their specific photosynthetic activities, $s_A F_{MAX}$, are highly correlated with the nitrogen contents of their leaves. In general the species with lower leaf nitrogen content have leaves with lower specific leaf areas (i.e. thicker leaves) or, more crudely, the plants with lower specific leaf photosynthetic activity have larger photosynthetic systems beneath unit leaf area.

In the first example the differences in specific leaf photosynthetic activity are environmentally induced, in the second they are genetically determined. In both cases the specific activity is related to the nitrogen concentration of the tissues.

In contrast to the *Eucalyptus* species, the genetic variability in F_{MAX} of most cultivated species appears to be associated with differences in the size of the photosynthetic apparatus beneath unit leaf area. For example, genetic differences in F_{MAX} have been correlated with the reciprocal of the specific leaf area in alfalfa, oats and soya bean, and with leaf thickness in soya bean, rye grass and sugar cane. Data for different genotypes of *L. perenne* are shown in Fig. 2.12. These correlations suggest that in the cultivated species a major source of the genetic variation in leaf photosynthetic activity lies in differences in the size of the photosynthetic apparatus beneath unit leaf area rather than in the specific or 'intrinsic' activity of that apparatus. The consequences of this observation are examined in Chapter 4.

2.11. CONCLUSIONS

It is relevant, indeed it is necessary, to provide a satisfactory answer to the question 'what contribution has the modelling exercise made to the understanding of the physiology of leaf photosynthesis?' Because the resolution of the separate physical and biochemical processes through analysis of photosynthesis measurements at the whole leaf level is poor the model does not help us make any definitive statement about these separate processes. However, by using mathematics to formalize our hypotheses based on their *in vitro* properties we have obtained a useful and robust mathematical description of the instantaneous response of leaf photosynthesis to changes in the leaf's environment. The realistic properties of this model also lend credence to both the validity of the original hypotheses and the reasonableness of our physiological definitions of the model's parameters.

By 'fitting' the model to experimental data we are able to obtain numerical estimates of the parameters of the model together with estimates of their approximate confidence intervals. We can then compare the numerical estimates obtained from the analysis of photosynthesis data obtained for different leaves and thereby make some sort of comparison of the photosynthetic behaviour of the leaves. Thus we have been able to demonstrate that the complex ratio $\beta K_o / \tau K_c$ does not appear to differ, at a particular temperature, between leaves from different C_3 plant species (see Table 2.2). Why the ratio does not change is not particularly important, but the observation that it does not is important. The observation implies that there is little, if any, variation in the ratio of photorespiratory to photosynthetic activity across widely different C_3 plant species. In an agricultural sense photorespiration is wasteful, the process consumes carbohydrate formed during photosynthesis, and it would be attractive to eliminate it through plant breeding. These data suggest that this is unlikely to be possible with existing varieties, since if there is little variation in the ratio between species there is unlikely to be any within species. The C_4 plants appear to have effectively suppressed photorespiration by evolving a mechanism by which carbon dioxide is 'concentrated' at the photosynthetic site. The same analysis enables us to estimate that in the leaves of C_4 plants the transport constant for movement of carbon dioxide to the photosynthetic site is about twenty times the transport constant for movement in the opposite direction.

There are sound experimental reasons for expressing photosynthetic rates on the basis of leaf area. Mathematical models of the biochemical processes are forced, however, to express rates on the basis of the volume of the photosynthetic apparatus since these reactions depend on enzyme and substrate concentrations, all functions of the volume of the system. In order that we can interface the experimental measurements with our theoretical model we are obliged to introduce a length dimension, leaf thickness. Arguably, the most important contribution of the modelling exercise has been the introduction of this dimension. It readily leads to us distinguishing two important properties of the leaf: first, its specific photosynthetic activity, i.e. its activity per unit leaf volume or weight, and second, the size of the photosynthetic apparatus beneath unit leaf area. We have been able to demonstrate that both environmental adaptation and genetic variation can be unambiguously attributed to differences in one or other of these properties. It has been demonstrated in Section 2.8 that changes in the size of the photosynthetic apparatus beneath unit leaf area in tomato are also associated with changes in the anatomical structure of the leaf which minimize the effects of leaf thickening on the physical processes of carbon dioxide exchange between the leaf and the air.

The model has helped us create some semblance of order from apparently disordered observations, and has thereby contributed usefully to our understanding of leaf photosynthesis. Also, it provides a simple, mathematical description of a complex phenomenon, and we will see in Chapter 3 that the simplification is useful in helping integrate our understanding of photosynthesis by the individual leaf into an understanding of photosynthesis and growth of a community of plants.

2.12. MAIN SYMBOLS

Some of the symbols are used throughout this chapter, and these are listed below. Others are used only in one or two sections, and are listed under the section where they are introduced.

Symbol	*Meaning*	*Unit*
C	ambient carbon dioxide concentration	$g\ m^{-3}$
O	ambient oxygen concentration	$g\ m^{-3}$
W	ambient water vapour concentration	$g\ m^{-3}$

The subscripts b and i refer to the inter-cellular and intra-cellular concentrations respectively.

I	light flux density incident on the leaf surface	$W\ m^{-2}$ or $J\ m^{-2}\ s^{-1}$
h	leaf thickness	m
P	net rate of photosynthesis per unit leaf volume	$g\ m^{-3}\ s^{-1}$
F	net rate of photosynthesis per unit leaf area	$g\ m^{-2}\ s^{-1}$

The subscript MAX (i.e. P_{MAX} or F_{MAX}) refers to the net rate of 'light saturated photosynthesis.

α_m	maximum light utilization efficiency	$g\ J^{-1}$
τ	maximum carboxylation efficiency	s^{-1}
β	maximum oxygenation efficiency	s^{-1}
Γ	ambient carbon dioxide concentration at the leaf compensation point (i.e. when P or $F = 0$)	$g\ m^{-3}$

Section 2.2

Q	light energy absorbed by the leaf	$W\ m^{-3}$ or $J\ m^{-3}\ s^{-1}$
P_G	gross rate of photosynthesis per unit leaf volume	$g\ m^{-3}\ s^{-1}$
R_L	photorespiration rate per unit leaf volume	$g\ m^{-3}\ s^{-1}$
R_D	'dark' respiration rate per unit leaf volume	$g\ m^{-3}\ s^{-1}$
δ	reciprocal of the maximum potential rate of gross photosynthesis	$g^{-1}\ m^3\ s$
γ	reciprocal of the maximum potential rate of photorespiration	$g^{-1}\ m^3\ s$

Section 2.3

T	transpiration rate per unit leaf volume	$g\ m^{-3}\ s^{-1}$
ϕ	diffusion constant	s^{-1}

The subscripts s, so and wo refer to the stomatal diffusion constants for carbon dioxide, oxygen and water vapour respectively.

k transport constant s^{-1}

The subscripts 1 and 10 refer to movement of carbon dioxide and oxygen from the intercellular air spaces to the photosynthetic sites and the subscripts 2 and 20 to their movement in the opposite direction.

Section 2.4

R_d 'dark' respiration rate per unit leaf area $g\ m^{-2}\ s^{-1}$
α actual light utilization efficiency $g\ J^{-1}$
K_o ratio of transport constants for oxygen (k_{10}/k_{20})

Section 2.5

ϕ_m mesophyll diffusion constant for carbon dioxide s^{-1}
K_c ratio of transport constants for carbon dioxide
        (k_1/k_2)

Section 2.6

T temperature $^\circ K$
R gas constant $cal\ mol^{-1}\ deg^{-1}$
E activation energy $cal\ mol^{-1}$
ϕ overall leaf carboxylation constant s^{-1}

Section 2.7

ψ_ℓ leaf water potential   bar
$\Delta_{\frac{1}{2}}$ stomatal 'sensitivity' to leaf water deficits $m\ s^{-1}\ bar^{-1}$
n constant

Section 2.8

p stomatal density (i.e. number per unit leaf area) m^{-2}
a mean cross sectional area of stomatal aperture m^2
A leaf area m^2
m mesophyll cell numbers beneath unit leaf area m^{-2}
r mean mesophyll cell radius m
ϱ mean mesophyll cell cross sectional area m^2
V leaf volume m^3
q mesophyll cell numbers in unit leaf volume m^{-3}
b cell 'packing' constant
D^* 'effective' diffusion coefficient $m\ s^{-1}$

Subscripts s and m refer to stomatal and mesophyll coefficients.

Section 2.9

s_A specific leaf area $m^2 \; g^{-1}$

2.13. SUGGESTED FURTHER READING

Section 2.2

Charles-Edwards, D. A. and Ludwig, L. J. (1974). *Ann. Bot.* **38**, 921–930.
Hall, A. E. (1971). *Carnegie Institution Year Book* **70**, 530–540.
Laisk, A. (1976). *In* 'Prediction and Measurement of Photosynthetic Activity'.
    Centre for Agricultural Publishing and Documentation, Wageningen.
Lorimer, G. H. and Andrews, T. J. (1973). *Nature, Lond.* **243**, 359–360.
Ogren, W. L. and Bowes, G. (1971). *Nature, New Biol.* **230**, 159–160.

Section 2.3

Bykov, O. D. and Levin, E. S. (1976). *Fiziologiya Rastenii* **23**, 238–246.
Bierhuizen, J. F. and Slatyer, R. O. (1965). *Agric. Meteorol.* **2**, 259–270.
Charles-Edwards, D. A. (1971). *Planta (Berl.)* **101**, 43–50.
Jarman, P. D. (1974). *J. exp. Bot.* **25**, 927–936.
Lake, J. V. (1967). *Aust. J. biol. Sci.* **20**, 487–493.

Section 2.4

Charles-Edwards, D. A. and Ludwig, L. J. (1974). *Ann. Bot.* **38**, 921–930.
Chartier, P. and Prioul, J. L. (1976). *Photosynthetica* **10**, 20–24.
Hall, A. E. and Bjorkman, O. (1975). *Ecological Studies* **12**, 55–71.
Peisker, M. and Apel, P. (1977). *Photosynthetica* **11**, 29–37.
Sinclair, T. R., Goudrian, J. and de Wit, C. T. (1977).*Photosynthetica* **11**, 56–65.
Tenhuen, J. D., Yocum, C. S. and Gates, D. M. (1976). *Oecologia (Berl.)* **26**,
    89–100.

Section 2.5

Charles-Edwards, D. A. (1978). *Ann. Bot.* **42**, 733–739.
Charles-Edwards, D. A. and Ludwig, L. J. (1975). *In* 'Environmental and Bio-
    logical Control of Photosynthesis' (R. Marcelle, ed.). Junk, The Hague.
Peisker, M. and Apel, P. (1976). *Photosynthetica* **10**, 140–146.
Tenhuen, J. D., Weber, J. A., Filipek, L. H. and Gates, D. M. (1977). *Oecologia*
    (Berl.) **30**, 189–207.

Section 2.6

Bjorkman, O. and Ehleringer, J. (1975). *Carnegie Institution Year Book* **74**,
    760–761.
Charles-Edwards, D. A. and Charles-Edwards, J. (1970). *Planta (Berl.)* **94**,
    140–151.
Charles-Edwards, D. A., Charles-Edwards, J. and Cooper, J. P. (1971). *J. exp.*
    Bot. **22**, 650–662.
Ehleringer, J. and Bjorkman, O. (1977). *Plant Physiol.* **59**, 86–90.

Section 2.7

Cowan, I. R. (1972). *Planta (Berl.)* **106**, 185–219.
Fisher, M. J., Charles-Edwards, D. A. and Ludlow, M. M. (1981). *Aust. J. Plant Physiol.* (in press).
McCree, K. J. (1974). *Crop Sci.* **14**, 273–278.

Section 2.8

Chabot, B. F., Junk, T. W. and Chabot, J. F. (1979). *Amer. J. Bot.* **66**, 940–945.
Charles-Edwards, D. A., Charles-Edwards, J. and Sant, F. I. (1972). *Planta (Berl.)* **104**, 297–305.
Charles-Edwards, D. A., Charles-Edwards, J. and Sant, F. I. (1974). *J. exp. Bot.* **25**, 715–724.
Wilson, D. (1975). *Ann. appl. Biol.* **79**, 83–95.
Wilson, D. and Cooper, J. P. (1969). *New Phytol.* **68**, 627–644; 645–655; 1115–1123; 1125–1135.

Section 2.9

Charles-Edwards, D. A. (1979). *In* 'Photosynthesis and Plant Development' (R. Marcelle, H. Clijsters and M. Van Pouke, eds). Junk, The Hague.
Charles-Edwards, D. A. and Ludwig, L. J. (1975). *In* 'Environmental and Biological Control of Photosynthesis' (R. Marcelle, ed.). Junk, The Hague.
Charles-Edwards, D. A. (1979). *Ann. Bot.* **44**, 523–525.
Nobel, P. S. (1977). *Physiol. plant.* **40**, 137–144.
Patterson, D. T., Duke, S. O. and Hoagland, R. E. (1978). *Plant Physiol.* **61**, 402–405.

Section 2.10

Mooney, K. A., Ferrar, P. J. and Slatyer, R. O. (1978). *Oecologia (Berl.)* **36**, 103–111.

Section 2.11.

Charles-Edwards, D. A. (1978). *Ann. Bot.* **42**, 717–731.
Thornley, J. H. M. (1976). *In* 'Mathematical Models in Plant Physiology.' Ch. 4. Academic Press, London and New York.

3. Canopy Photosynthesis

3.1. INTRODUCTION

It is logical to attempt to extend the model for leaf net photosynthesis, described in Chapter 2, to deal with the photosynthetic behaviour of a community of plants. The mathematical description of leaf photosynthesis is dominated by the kinetic behaviour of the biochemical processes, and the apparent 'competition' between carbon dioxide and oxygen for the same acceptor molecule, ribulose-1, 5-bisphosphate. It should be apparent that, however carefully made, experimental measurements of the net rates of carbon dioxide exchange by intact leaves do not allow the unambiguous identification and characterization of the subtle mathematical differences arising from different assumptions about the detail of the physical and biochemical processes of photosynthesis. It is unlikely that an analysis of canopy photosynthesis data will allow the full detail of the individual component leaf's behaviour to be resolved. The mathematical description of canopy photosynthesis is most likely to be dominated by the hyperbolic response of the rate of leaf net photosynthesis to increasing light flux densities and ambient carbon dioxide concentrations. Accordingly, it is assumed at the outset that the rate of leaf net photosynthesis per unit leaf area, F, is adequately described by

$$F = \alpha_m I(\tau C - \beta) / (\alpha_m I + \tau C) - R_d, \qquad (3.1)$$

where I is the incident light flux density, C is the ambient carbon dioxide concentration, R_d is the rate of 'dark' respiration per unit leaf area, α_m is the light utilization efficiency at high ambient carbon dioxide concentrations, τ is an 'overall' leaf carboxylation conductance (including a component due to the stomatal conductance) and β is a constant for photorespiration in the in the leaves of C_3 plants. For canopies consisting of leaves of C_4 plants β will be zero. Indeed, it may not be possible to make any meaningful estimate of β from measurements of canopy net photosynthetic rates, and eqn (3.1) may be reasonably reduced to

$$F = \alpha I \tau C / (\alpha I + \tau C) - R_d, \qquad (3.2)$$

where α is the leaf light utilization efficiency, for canopies consisting of leaves of both C_3 and C_4 plant species.

The major problem encountered in modelling canopy net photosynthetic rates in terms of leaf photosynthetic behaviour is the definition of the separate environments of the individual leaves which constitute the canopy. The simplest situation is that of the 'closed crop'. The term 'closed crop' defines a community of plants, of uniform height, which extends indefinitely in the horizontal plane. Within a 'closed crop' canopy we might expect the leaves in any particular horizon to experience a uniform environment, and we might further expect the only significant source of environmental variation to be found in the vertical plane. An assumption often implicit in the term 'closed crop' is that leaf area density (leaf area per unit volume of canopy) is constant throughout the canopy.

The development and applications of mathematical models for photosynthesis by 'closed crop' canopies are described in Sections 3.2.–3.4. A model based on the assumption of uniform leaf photosynthetic characteristics throughout the canopy is constructed in Section 3.2. The model has been used to analyse canopy photosynthesis data, and these analyses are discussed in Section 3.3. There are data which indicate that leaf photosynthetic characteristics are not uniform throughout the canopy, but may vary in a simple and systematic way. Two models in which the assumption of uniform leaf photosynthetic character- istics is relaxed are constructed in Section 3.4. Their behaviour is examined and one of them is used to analyse experimental data.

The rate of canopy net photosynthesis includes a component due to the canopy 'dark' respiration rate. The rate of 'dark' respiration by a community of plants includes contributions from both the photosynthetically active tissues (primarily the leaves) and the non-photosynthetically active tissues such as roots and fruits. Section 3.5 deals with some of the problems encountered in modelling canopy 'dark' respiration rates.

Many crops during the whole period of their growth, and all crops during the early stages of growth, cannot be treated as 'closed crops'. For example, many horticultural crops are grown as row crops—field crops, such as maize, grow as row crops during their early growth, whilst tree crops, such as apple orchards, have discrete leaf canopies. In all these examples there are considerable variations in the downward light flux density in both the horizontal and vertical planes. An important problem in modelling canopy photosynthesis in these crops is the prediction of the light environments of leaves within the canopy. An approach to this problem is described in Section 3.6. The mathematical description of the light environment can be combined with suitable descriptions of leaf net photosynthesis and then used to predict the rate of canopy net photosynthesis. Some predictions are compared with experimental observations in Section 3.7.

3.2. A SIMPLE MODEL

The major source of environmental variation within a 'closed crop' canopy can be attributed to the deteriorating light environment of the separate leaves with their increasing depth below the uppermost surface of the canopy. It is not unreasonable to suppose, as a first approximation, that the ambient carbon dioxide concentration remains fairly constant throughout the volume of the canopy. With many 'closed crops' the downward light flux density through a horizontal plane, I_H, appears to decrease exponentially with increasing depth below the uppermost surface of the canopy. The change in downward light flux density is simply, and usefully, described by Beer's Law. If we denote the downward light flux density immediately above the canopy by I_0, we can write that below a cumulative leaf area index L,

$$I_H = I_0 \exp(-KL), \tag{3.3}$$

where K is a constant called the canopy extinction coefficient.

Now the downward light flux density through a horizontal plane may not be the same as the light flux density incident on a leaf's surface in that plane and available for photosynthesis. If we consider the downward component of light over an element of leaf area index ΔL, there will be a reduction in I_H of ΔI_H as light is absorbed and scattered as it passes through the element of leaf area index. If we neglect the scattered light we can write

$$-\Delta I_H = (1 - m)I_H \, \Delta L, \tag{3.4}$$

where m is the proportion of the incident light transmitted by the leaves. The negative sign in front of ΔI_H shows that there is a *decrease* in the downward light flux density, that is light is *absorbed* by the leaves. If we re-arrange eqn (3.4), replacing I_H by the symbol I, and take the limit as that $\Delta I_H/\Delta L \to dI_H/dL$, the light incident on the leaves and available for photosynthesis becomes

$$I = -(dI_H/dL) / (1 - m) \tag{3.5}$$

and from eqn (3.3) we have

$$dI_H/dL = -KI_0 \exp(-KL). \tag{3.6}$$

If we substitute for dI_H/dL in eqn (3.5) it is a simple matter to obtain the expression

$$I = KI_0 \exp(-KL)/(1 - m), \tag{3.7}$$

which describes the light flux density incident on the leaf surface and available for photosynthesis as a function of the downward light flux density at the top of the 'closed canopy', the canopy extinction coefficient, the leaf light transmission coefficient and the cumulative leaf area index above the particular horizontal plane of interest.

With a row crop, or isolated plant canopy such as the canopy of an individual tree in full leaf, there will be marked spatial variation in the downward light flux density within the canopy in both the horizontal and the vertical planes. One approach to the analysis of this problem is outlined in Section 3.6.

A mathematical description of the rate of net photosynthesis per unit of ground area by a 'closed crop' canopy can be obtained by combining eqns (3.1) and (3.7) and then integrating the consequent expression over the entire leaf area of the canopy. We have

$$F = \frac{\alpha_m (\tau C - \beta) K I_0 [\exp(-KL)]}{\alpha_m K I_0 \exp(-KL) + (1 - m)\tau C} - R_d \qquad (3.8a)$$

and

$$F_c = \int_0^L F \, dL, \qquad (3.8b)$$

where F_c is the rate of net photosynthesis per unit of ground area by the 'closed crop' canopy. Combination of eqns (3.8a) and (3.8b) yields

$$F_c = \frac{\tau C - \beta}{K} \ln \left[\frac{\alpha_m K I_0 + (1 - m)\tau C}{\alpha_m K I_0 \exp(-KL) + (1 - m)\tau C} \right] - R_c, \qquad (3.9)$$

where R_c denotes the rate of canopy 'dark' respiration per unit of ground area. The term R_c includes a component due to the non-photosynthetic tissues (stems, roots and fruits etc.), and is not simply given by $R_d L$ as would be strictly inferred by eqns (3.8a) and (3.8b). It should be noted that we have implicitly assumed that the photosynthetic characteristics of the individual leaves within the 'closed crop' canopy are identical, that is α_m, β and τ do not vary with leaf position.

Whilst eqn (3.9) appears a little cumbersome mathematically, it provides a a fairly simple description of the instantaneous response of the rate of canopy net photosynthesis to changes in the downward light flux density incident at the top of the canopy and the ambient carbon dioxide concentration. Two properties of eqn (3.9) are worth further examination. First, if the incident light flux density increases until the rate of canopy photosynthesis becomes 'light saturated' we can write this rate, $(F_c)_{MAX}$, as

$$(F_c)_{MAX} = (\tau C - \beta)L - R_c, \qquad (3.10)$$

since $\ln [\quad]$ in eqn (3.9) tends to KL. That is, the rate of 'light saturated' canopy net photosynthesis is linearly dependent of the leaf area index of the crop. Second, if we differentiate F_c with respect to I_0, and take the limit as $I_0 \to \infty$, the canopy light utilization efficiency, α_c, is given as

$$\alpha_c = \left[\frac{dF_c}{dI_0} \right]_{I_0 \to \infty} = \frac{\alpha_m (\tau C - \beta)[1 - \exp(-KL)]}{(1 - m)\tau C}. \qquad (3.11)$$

We can re-arrange eqn (3.11) to give

$$(1 - m)\alpha_c / [1 - \exp(-KL)] = \alpha_m (1 - \beta/\tau C), \qquad (3.12)$$

which is directly comparable with eqn (2.37). Equation (2.37) defines the light utilization efficiency of a single leaf of a C_3 plant. Whereas eqn (3.10) suggests that $(F_c)_{MAX}$ is linearly dependent on L, eqn (3.11) indicates that α_c will show a maximum value as L increases and the term $[1 - \exp(-KL)]$ approaches unity.

3.3. AN ANALYSIS OF CANOPY PHOTOSYNTHESIS DATA

Since eqn (3.9) describes the instantaneous response of the rate of canopy net photosynthesis to changes in the downward light flux density incident on the top surface of the 'closed crop' canopy, it provides a useful analysis for the photosynthetic response of a crop to the predictable diurnal variation in incident light experienced in the field. Superimposed on the systematic variation will be random fluctuation due to the natural minute-to-minute variation in both cloud density and the degree of cloudiness of the sky. Equation (3.9) also describes the effects of changes in the ambient carbon dioxide concentration on canopy photosynthesis. Although the ambient carbon dioxide concentration does not change greatly during the course of the day, and from day to day, out of doors, the glasshouse crop growers in northern Europe commonly enrich the carbon dioxide atmosphere of their glasshouses. Whereas the natural out of doors concentration is about 0.6 g (CO_2) m^{-3}, growers commonly enrich glasshouses, during the winter months when the glasshouse ventilator panels are closed, to about 2.0 g (CO_2) m^{-3}. Equation (3.9) provides a useful tool for investigating the physiological effects of this carbon dioxide enrichment.

For example, the responses of the rates of canopy net photosynthesis of 'closed crops' of tomato (*Lycopersicon esculentum*) to incident light flux density at two ambient carbon dioxide concentrations are shown in Fig. 3.1. The responses are linear over the range of incident light flux densities 0-60 W m^{-2}. The canopy light utilization efficiencies for these crops can be obtained as the slopes of the linear regression of F_c on I_0. Values of α_c, so obtained, for 'closed crops' of tomato plants photosynthesizing at different ambient carbon dioxide concentrations and water vapour deficits are given in Table 3.1. The light utilization efficiency of the 'closed crop' increases with increasing ambient carbon dioxide concentrations and, in general, dec-

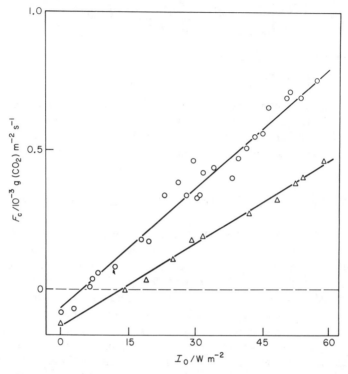

FIG. 3.1. Canopy photosynthesis vs. incident light flux density response of 'closed crops' of tomato (*Lycopersicon esculentum*) at two ambient carbon dioxide concentrations. Solid lines represent the linear regression of F_c on I_0. (○, 2.20 g (CO_2) m^{-3}; △, 0.73 g (CO_2) m^{-3}.)

TABLE 3.1. Mean values of the ambient carbon dioxide concentration, ambient vapour pressure deficit (vpd) and canopy light utilization coefficient for a 'closed crop' of tomato plants (*Lycopersicon esculentum*). In all cases $[1 - \exp(-KL)]/(1 - m)$ was 1.06.

C/g (CO_2) m^{-3}	vpd/kPa	$\alpha_c/10^{-6}$ g (CO_2) J^{-1}
0.73	0.46	10.4
1.46	0.51	12.4
2.20	0.54	13.5
0.73	0.79	10.2
1.46	0.78	11.0
2.20	0.84	11.7
0.73	1.05	9.9
1.46	1.09	11.4
2.20	1.06	13.5

reases with increasing water vapour deficits. Even at low incident light flux densities, therefore, carbon dioxide enrichment of the glasshouse, particularly when combined with low water vapour pressure deficits, is likely to enhance crop photosynthesis, and presumably total crop dry matter yield.

In the previous section we were able to show that

$$(1 - m)\alpha_c/[1 - \exp(-KL)] = \alpha_m(1 - \beta/\tau C). \qquad (3.12)$$

If we denote $(1 - m)\alpha_c/[1 - \exp(-KL)]$ by α^*, eqn (3.12) predicts that α^* will be directly proportional to the reciprocal of the ambient carbon dioxide concentration. The data given in Table 3.1 provide support for this prediction (see Fig. 3.2). If we also assume that τ, the overall leaf carboxylation conductance, is inversely proportional to the water vapour pressure deficit, then α^* should be proportional to the vapour pressure deficit divided by the ambient carbon dioxide concentration. The relationship between α^* and vpd/C is shown in Fig 3.2. The assumption that $\tau \propto 1/$vpd follows from that fact that a

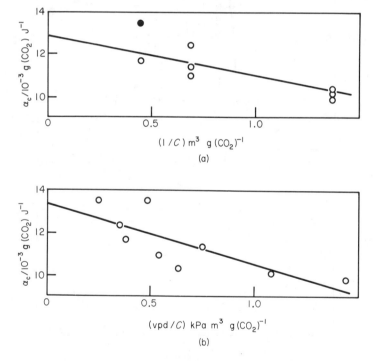

FIG. 3.2. (a) The relationship between the estimated leaf light utilization efficiency, α, and the reciprocal of the ambient, carbon dioxide concentration (C). (b) The relationship between the estimated light utilization efficiency and the ratio of the ambient water vapour pressure deficit and carbon dioxide concentration, vpd/C.

component of the overall leaf carboxylation conductance is the stomatal conductance. As the water vapour deficit in the bulk air surrounding the leaf increases it is reasonable to suppose that the leaf transpiration rate will also increase. To minimize the loss of water by the leaves we might expect the stomates to close, with an associated decrease in the stomatal conductance. If the stomatal conductance decreases sufficiently there will be a decrease in the overall leaf carboxylation conductance. The data shown in Fig. 3.2 are in accord with this hypothesis.

Typical changes in the instantaneous rate of canopy photosynthesis,

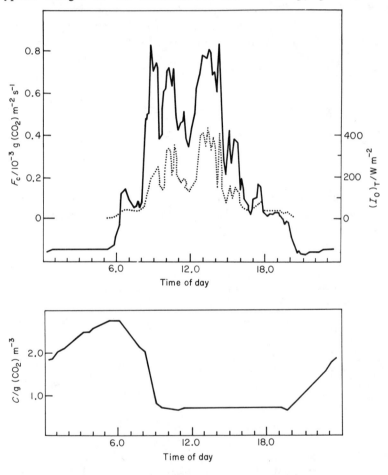

FIG. 3.3. A typical diurnal response of canopy net photosynthetic rate, F_c, to ambient carbon dioxide concentration, C, and incident radiation flux density, I_0, for a 'closed crop' of chrysanthemum plants growing in a daylight assimilation chamber.

attributable to the natural diurnal variation in incident radiation flux density, for a 'closed crop' of vegetative plants of *Chrysanthemum morifolium* in natural daylight, are shown in Fig. 3.3. Shown also in the figure are the measured ambient carbon dioxide concentrations and incident radiation flux densities. Chrysanthemum crops were grown for 6–8 week periods in daylight assimilation chambers. The sides of the crops were screened with green netting to simulate 'closed crop' conditions. The instantaneous rates of canopy photosynthesis were measured by recording over successive 5 or 10 min intervals the amounts of carbon dioxide 'injected' into the assimilation chambers to replace the ambient carbon dioxide within the chambers that had been 'fixed' by the crop(s). The average incident radiation flux densities and ambient carbon dioxide concentrations experienced by the crop during the 5 or 10 min intervals were also recorded. Whilst ambient carbon dioxide concentrations within the chambers could be kept reasonably constant when the crops were photosynthesizing (removing the carbon dioxide), it was not possible to maintain any control over the concentrations when canopies were respiring in the dark (expiring carbon dioxide). This led to the considerable increase in ambient carbon dioxide concentration at night shown in Fig 3.3. In all the experiments with chrysanthemum crops reported in this section, *total radiation* flux densities were measured, not just the photosynthetically active component of the incident radiation flux density. To emphasize this point the symbols α_m, K, m and I_0 are re-written as $(\alpha_m)_T$, K_T, m_T and $(I_0)_T$ in all analyses of these data. Data of the sort shown in Fig. 3.3 can be analysed using eqn (3.9) to obtain estimates of the 'leaf parameters' $(\alpha_m)_T$, τ and β, and to investigate how these leaf characteristics change in response to long term changes in the crop's growth environment. The other parameters of eqn (3.9), K_T, m_T and L, and the variable R_c, are all amenable to direct measurement.

Equation (3.9) can be 'fitted' to data from the diurnal response of canopy net photosynthesis rate with the ambient carbon dioxide concentration and incident radiation flux density (C and $(I_0)_T$) being used as the two independent variables. The correlation between the 'fitted' and actual observed values of F_c for two such sets of data are shown in Fig. 3.4. The 'fitted' values are scattered about the 45° regression line, the expected line for 'perfect fit', and there is no consistent departure of the 'fitted' values from this line. We can have some confidence that the model provides a good qualitative description of these data.

The numerical estimates of the parameters $(\alpha_m)_T$, τ and β, obtained by 'fitting' eqn (3.9) to experimental data of the sort shown in Fig. 3.3, and the measured values of the canopy leaf area index, L, and dark respiration rate per unit ground area, R_c, are given in Table 3.2. Although the measured leaf area index increases from 2.5 to 8.2 there are no associated changes in $(\alpha_m)_T$, τ, β or R_c. The radiation utilization efficiency, $(\alpha_m)_T$, is primarily

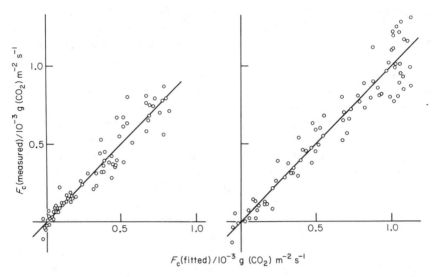

FIG. 3.4. The relationship between 'fitted' (using eqn (3.9)) and observed values of the canopy net photosynthetic rate of a 'closed crop' of chrysanthemum plants on separate occasions. The solid line is the expected line of 'perfect fit'.

determined by the canopy net photosynthetic rates measured at low incident radiation flux densities. At these low incident radiation levels approximately 60% of the radiation energy is in the photosynthetically active wavebands. The chrysanthemum canopy is also more 'transparent' to the non-photosynthetically active radiation, and K_T is about 30% less than the value of K for the photosynthetically active portion of the incident radiation alone. We can calculate that the values of $(\alpha_m)_T$ shown in Table 3.2 will be about one-half to two-thirds of the values of the light utilization efficiencies, α_m, of the leaves. It is also apparent from Table 3.2 that β is highly correlated with the parameter τ, and this correlation appears across data sets, as is demonstrated in Fig. 3.5, where the numerical values of β and τ shown in Table 3.2 are plotted against each other. Although chrysanthemum is a C_3 plant and its leaves exhibit photorespiration, the resolution of the leaf processes at the level of these canopy data is insufficient to enable a precise and unambiguous estimate of β. The correlation between the numerical estimates of β and τ can be mitigated by deriving another leaf photosynthetic characteristic from them which subsumes their correlation. The rate of light saturated photosynthesis per unit leaf area, F_{MAX}, can be written as

$$F_{MAX} = (\tau C - \beta) - R_d, \qquad (3.13)$$

where, as a first approximation, we can assume that $R_d = R_c/L$.

TABLE 3.2. Values of the parameters $(\alpha_m)_T$, τ and β, together with their approximate 95% confidence intervals, obtained by fitting eqn (3.9) to canopy photosynthesis data for a chrysanthemum crop at different times during its growth. L and R_c are measured leaf area indices and canopy dark respiration rates.

L	$(\alpha_m)_T/\text{kg}\,(CO_2)\,\text{J}^{-1}$	$\tau/\text{m s}^{-1}$	$\beta/\text{kg}\,(CO_2)\,\text{m}^{-2}\,\text{s}^{-1}$	$R_c/\text{kg}\,(CO_2)\,\text{m}^{-2}\,\text{s}^{-1}$
2.5	$1.3\,(0.2) \times 10^{-8}$	$1.12\,(0.21) \times 10^{-3}$	$1.6\,(1.0) \times 10^{-7}$	2.1×10^{-7}
3.8	$1.2\,(0.2) \times 10^{-8}$	$0.95\,(0.17) \times 10^{-3}$	$1.2\,(0.9) \times 10^{-7}$	1.8×10^{-7}
4.0	$1.2\,(0.2) \times 10^{-8}$	$0.98\,(0.20) \times 10^{-3}$	$0.9\,(1.1) \times 10^{-7}$	2.3×10^{-7}
5.6	$1.0\,(0.1) \times 10^{-8}$	$1.91\,(0.53) \times 10^{-3}$	$4.0\,(4.0) \times 10^{-7}$	1.7×10^{-7}
7.2	$1.5\,(0.3) \times 10^{-8}$	$0.75\,(0.18) \times 10^{-3}$	$-0.1\,(0.8) \times 10^{-7}$	3.3×10^{-7}
8.2	$1.1\,(0.1) \times 10^{-8}$	$1.76\,(0.42) \times 10^{-3}$	$2.0\,(1.2) \times 10^{-7}$	2.0×10^{-7}

'Closed crops' of vegetative chrysanthemum plants have been grown in daylit assimilation chambers at different times of year and at the same time of year but

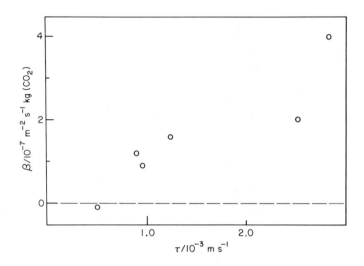

FIG. 3.5. The observed relationship between the numerical estimates of β and τ obtained by 'fitting' eqn (3.9) to six separate sets of data on the canopy net photosynthetic rate of a 'closed crop' of chrysanthemum plants.

TABLE 3.3. The effects of average daily total radiation integral, S^*, and growth temperature, T, on leaf radiation utilization efficiency, α_m^*, the rate of 'light saturated' leaf photosynthesis, F_{MAX}, and specific leaf area of 'average' leaves in a chrysanthemum crop.

$S^*/\text{MJ m}^{-2}$	$T/^\circ\text{C}$	$\alpha_m^*/\text{kg (CO}_2)\,\text{J}^{-1}$	$F_{MAX}/10^{-6}\text{kg(CO}_2)\,\text{m}^{-2}\text{s}^{-1}$	$(1/s_A)/\text{m}^2\,\text{kg}^{-1}$
1.9	20	1.1×10^{-8}	0.27	44
4.1	20	1.0×10^{-8}	0.43	33
5.4	20	1.0×10^{-8}	0.49	27
8.5	20	0.9×10^{-8}	0.56	27
9.2	20	1.2×10^{-8}	0.56	24
5.4	10	0.9×10^{-8}	0.65	16
5.4	15	0.9×10^{-8}	0.58	24
5.4	20	1.0×10^{-8}	0.49	27
5.4	25	1.0×10^{-8}	0.59	29
5.4	30	0.9×10^{-8}	0.61	29

subject to different bulk air temperatures. Data of the type shown in Fig. 3.3 were collected at regular intervals during the growth of each of these crops. The data have been analysed using eqns (3.9) and (3.13), and mean values of $(\alpha_m)_T$ and F_{MAX} for each of the crops are shown in Table 3.3. The incremental specific leaf area, s_A, was also measured for each of the crops as the slope of the linear regression of leaf area on leaf dry weight for all plants harvested from each crop. The value of s_A for each of the crops is also shown in Table 3.3. There are no consistent differences in $(\alpha_m)_T$, the leaf radiation utilization efficiency, with crop growth environment. Whereas F_{MAX} increased in a hyperbolic way with increasing average daily radiation integrals during growth, it demonstrated a far

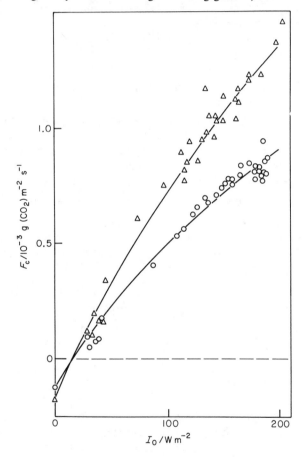

FIG. 3.6. The response of an intact ($\triangle$) and partially defoliated ($\bigcirc$) 'closed crop' of tomato (*Lycopersicon esculentum*) to changing incident light flux density. The solid lines were obtained by 'fitting' eqn (3.9) (with β put equal to 0) to each data set separately.

more complex response to changes in crop bulk air temperature. Specific leaf area declined with increasing daily radiation integrals and increased with increasing growth temperatures. The specific leaf area provides a crude estimate of leaf thickness, an increase in s_A is roughly equivalent to a decrease in leaf thickness. The rate of light saturated net photosynthesis per unit leaf dry weight, $F_{MAX}s_A$, remains fairly constant with changes in the average daily radiation integral but increases hyperbolically with increasing bulk air temperatures. These data are in accord with those discussed in Section 2.9. It is reassuring that although the leaf photosynthetic characteristics $(\alpha_m)_T$ and F_{MAX} are 'average' values for all leaves in the canopy, these 'average' values show similar adaptive responses to changing growth light and temperature regimes to the values obtained for single, identifiable leaves.

We have seen that β is only poorly determined by analysis of canopy photosynthesis data of the sort described here. Even if β is put equal to zero (i.e. we use eqn (3.2) to describe the rate of leaf net photosynthesis by C_3 plants), eqn (3.9) can still provide a useful qualitative description of canopy photosynthesis data. Equation (3.9), with β put equal to zero, has been used to analyse canopy photosynthesis data for a 'closed crop' of tomato plants (*L. esculentum*) before and after partial defoliation. The crop was defoliated from the base upwards to leave only the uppermost 25% of the normal leaf area. The experimental response of canopy net photosynthetic rate to changing incident light flux density and 'fitted' response curves are shown in Fig. 3.6. The estimated values of α_m and τ, together with the measured values of K, m, R_c and L, are given in Table 3.4.

TABLE 3.4. Measured values of K and L together with estimates of the leaf photosynthesis parameters α_m and τ for an intact 'closed crop' of tomato plants, and the same crop following partial defoliation from the base upwards. The numbers in brackets are the approximate 95% confidence intervals of α_m and τ.

	K	L	$\alpha_m/10^{-6}$ g (CO_2) J^{-1}	$\tau/10^{-3}$ m s^{-1}
Intact crop	0.52	8.6	9.6 (1.3)	1.6 (0.7)
Partially defoliated crop	0.63	2.0	9.1 (1.2)	2.1 (0.6)

3.4. AN EXTENSION TO THE MODEL

An important assumption in setting up the simple model described in Sections 3.2 and 3.3 is that the photosynthetic characteristics of all the leaves in the 'closed crop' canopy are the same. We have seen in Section 3.3 and in Section 2.9 that the light environment during growth and development affect a leaf's photosynthetic characteristics. It remains to be established whether long-term changes in the leaf's environment, after full leaf expansion, affect its photosynthetic performance. For example, in many crops new leaves develop at the top of the canopy, experiencing almost the same light flux density as is incident at the top surface of the canopy. However, the growth and development of subsequent new leaves above them leads to increasing shading of them. It remains to be established whether or not this shading affects their rates of photosynthesis.

Equation (3.2) has been 'fitted' to the net photosynthetic rate vs. incident light flux density response curves for individual leaves from different horizons within a stand of tomato plants (*L. esculentum*). The numerical estimates of the leaf photosynthetic parameters α and τ obtained from these analyses are given in Table 3.5. It is important to remember that eqn (3.2)

TABLE 3.5. Numerical estimates of the parameters α and τ obtained by 'fitting' eqn (3.2) to leaf photosynthesis data obtained for leaves from different horizons within a stand of tomato plants (*Lycopersicon esculentum*).

Leaf number	$\alpha/10^{-6}$ g(CO_2) J^{-1}	$\tau/10^{-3}$ m s^{-1}	Note
20	10.4 ($\pm$0.9)	1.7 ($\pm$0.2)	a
20	11.4 ($\pm$1.2)	1.8 ($\pm$0.1)	a
20	14.1 ($\pm$1.8)	2.1 ($\pm$0.2)	a
20	13.1 ($\pm$2.9)	0.6 ($\pm$0.1)	b
12	12.2 ($\pm$3.0)	0.4 ($\pm$0.1)	b
9	10.0 ($\pm$4.0)	0.2 ($\pm$0.1)	b
9	15.6 ($\pm$2.6)	1.1 ($\pm$0.1)	c

Note: a = unshaded leaf, b = shaded leaf, c = outer leaf adjacent to screen netting.

provides only an approximate description of single leaf photosynthesis, and generally leads to an overestimate of α and τ when 'fitted' to single leaf photosynthesis data. Despite this deficiency, it is patently obvious from Table 3.5 that there are considerable differences in the overall leaf carboxylation conductance, τ, for leaves from different positions within the tomato canopy. Leaves from the more heavily shaded positions at the base of the canopy have carboxylation conductances which are significantly lower than those for less shaded leaves at the top of the canopy. The differences are not attributable to leaf age since a young shaded leaf (leaf number 20) has a smaller carboxylation conductance than an older, partially-shaded leaf from the edge of the canopy (leaf number 9). This older leaf also has a larger overall carboxylation conductance than a leaf of the same age from a heavily shaded position in the centre of the canopy. In contrast to the leaf carboxylation conductance, there appears to be little consistent variation in the leaf light utilization efficiency α. It seems reasonable to suppose that the reduction in leaf conductance is attributable to the deteriorating light environment experienced by leaves with increasing depth in the canopy.

Equation (3.3) allows us to write that the degree of shading beneath a cumulative leaf area index L, I_H/I_0, is

$$I_H/I_0 = \exp(-KL), \tag{3.14}$$

and if we assume that the leaf overall carboxylation conductance is decreased in direct proportion to the degree of shading we can write

$$\tau = \tau_0 \exp(-KL), \tag{3.15}$$

where τ_0 is the conductance of a fully expanded, unshaded leaf at the top of the canopy. Substitution of eqns (3.7) and (3.15) into eqn (3.2) give the instantaneous rate of net photosynthesis for a horizon within the canopy beneath a cumulative leaf area index L as

$$F_L = \frac{\alpha I_0 K \tau_0 C \exp(-KL)}{\alpha I_0 K + (1 - m)\tau_0 C} - R_L, \tag{3.16}$$

where R_L is the dark respiration rate of that horizon. We can now write that the instantaneous rate of canopy net photosynthesis, F_c, will be given by

$$F_c = \int_0^L F_L \, dL, \tag{3.17}$$

and integration of eqn (3.16) leads to

$$F_c = \frac{\alpha I_0 \tau_0 C[1 - \exp(-KL)]}{\alpha I_0 K + (1 - m)\tau_0 C} - R_c, \tag{3.18}$$

where R_c is the rate of canopy 'dark' respiration per unit of ground area.

Equation (3.18) has been used to simulate the response of canopy net photosynthetic rate to changing incident light flux density for an intact and partially defoliated 'closed' crop of tomato plants. The experimental data have been described in the previous Section, and the 'fit' of eqn (3.9) to them has been illustrated in Fig. 3.6. Simulations of the responses, using eqn (3.18) and the parameter values shown in Table 3.6 are illustrated in Fig. 3.7. There is quite good agreement between the simulated and observed responses. Numerical values of L and K derive from experimental measurements of the canopies, and values of α, τ_0 and m from measurements of the constituent leaves.

Equation (3.18) has the important virtue of simplicity, and yet it enables us to predict, with reasonable accuracy, the photosynthetic behaviour of a 'closed crop' canopy from measurements of the photosynthetic characteristics of upper,

TABLE 3.6. Parameter values used to simulate the canopy photosynthesis response to light shown in Fig. 3.7.

α	11.5×10^{-6} g (CO_2) J^{-1}
τ_0	2.4×10^{-3} m s^{-1}
K	0.52
m	0.15

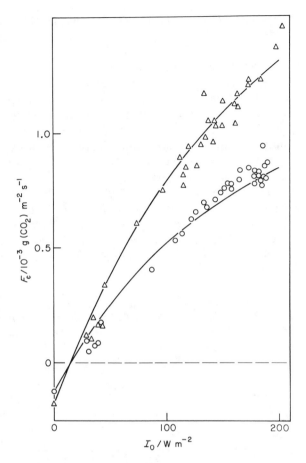

FIG. 3.7. The response of an intact ($\triangle$) and partially defoliated ($\circ$) 'closed crop' of tomato (*Lycopersicon esculentum*) to changing incident light flux density. The solid lines are simulations of the responses obtained by using eqn (3.18) with the parameter values given in Table 3.6.

unshaded leaves in the canopy. Another attraction of eqn (3.18) is that it can be re-arranged to

$$\frac{1-\exp(-KL)}{F_c+R_c} = \frac{K}{\tau_0 C} + \frac{1-m}{\alpha I_0}, \qquad (3.19)$$

which effectively 'linearizes' the response of canopy photosynthetic rate to changing ambient carbon dioxide concentrations and incident light flux densities. For example, $1 - \exp(-KL)$, which is the amount of incident light 'absorbed' by the 'closed crop' is readily measured, and a plot of $[1 - \exp(-KL)]/(F_c + R_c)$

should be linear with respect to the reciprocal of the incident light flux density at the top of the canopy (i.e. $1/I_0$). The slope will be given by $(1 - m)/\alpha$ and the intercept at $1/I_0 = 0$ by $K/\tau_0 C$. The parameters m and K and the independent variable C are also amenable to simple measurement.

The C_4 forage grass *Setaria anceps* has been grown as a 'closed sward', and the response of net photosynthesis by the sward to changing incident light flux densities measured. Prior to the photosynthesis measurements some swards were treated with giberellic acid to cause shoot elongation and create a less dense leaf canopy whilst other swards were treated with cyclocel to reduce shoot elongation and create a more dense leaf canopy. The objective of the experiment was to establish whether or not leaf density would affect the potential photosynthetic performance of the sward. The canopy photosynthesis data have been analysed using eqn (3.19). The estimated values of K/τ_0 and measured values of K and specific leaf area, s_A, for three crops of setaria (actually the mean values for swards treated with GA, CCC and untreated, control swards) are shown in Table 3.7. The analysis suggests that there is little effect of the different treatments on these characters.

TABLE 3.7. Numerical estimates of the canopy photosynthesis parameters K and τ_0, and the specific leaf area, s_A, obtained by analysis of data for the C_4 grass *Setaria anceps* using eqn (3.19). Swards were untreated and examined following spraying with the growth agents CCC (to dwarf growth) and GA (to cause shoot extension).

Treatment	K	$\tau_0/10^{-3}$ m s^{-1}	s_A/m^2 g^{-1}	$s_A\tau_0/10^{-5}$ m^3 g^{-1}s^{-1}
Control	0.58	4.6	0.018	8.3
CCC	0.61	4.6	0.017	7.8
GA	0.73	5.0	0.016	8.0

The grass has also been grown in mixed swards in association with the C_3 forage legume *Desmodium intortum*. The swards were regularly defoliated to two different heights and to each height at two different cutting frequencies. As each sward re-grew after it had been defoliated the diurnal variation in canopy net photosynthetic rate was measured. These data were analysed using eqn (3.19). Experimental measurements indicated that both the setaria and desmodium leaves were uniformly distributed throughout the canopy, and it was assumed that the average leaf carboxylation conductance for any leaf layer could be written as

$$\tau = x\tau_G + (1 - x)\tau_L, \qquad (3.20)$$

where τ_G and τ_L are the conductances of the grass and legume leaves and x is the proportion of grass leaf in any leaf layer.

Once the swards had recovered from the immediate effects of defoliation, generally within 2 or 3 d, their changing rates of canopy net photosynthesis could be attributed entirely to the increasing proportion of the incident radiation that they each intercepted as they regrew. Mean values of K, α and τ_0 for each sward during regrowth are shown in Table 3.8. The values of τ_0 for the 3 week cutting regimes were lower than those for the 5 week regimes, although the leaf light utilization efficiencies did not change.

TABLE 3.8. Numerical estimates of the canopy photosynthesis parameters K, α and τ_0 for mixed swards of the legume *Desmodium intortum* and the C_4 grass *Setaria anceps* obtained using eqn (3.19). Swards were cut at two frequencies (every 3 or 5 weeks) at two heights (7.5 or 15 cm above ground level).

Treatment	K	$1 - \exp(-KL)$	$\alpha/10^{-6}\ \text{g}\,(CO_2)\,J^{-1}$ [a]	$\tau_0/10^{-3}\ \text{m}\,\text{s}^{-1}$
3 week/7.5 cm	0.66	60–95%	8.1 (3.2)	6.6 (1.2)
3 week/15 cm	0.78	78–96%	8.3 (4.1)	5.8 (1.1)
5 week/7.5 cm	0.64	33–95%	8.4 (5.7)	8.8 (2.4)
5 week/15 cm	0.63	74–96%	8.3 (3.2)	8.8 (1.2)

[a] α is calculated from eqn (3.19) assuming a value of $m = 0.1$.

3.5. CANOPY RESPIRATION

It has frequently been argued that the canopy 'dark' respiration rate can be written as

$$R_c = a(dW/dt) + bW, \qquad (3.21)$$

where dW/dt is the crop growth rate, W the standing crop mass and a and b are constants. If the crop growth rate is assumed to be proportional to the 'average' rate of canopy net photosynthesis we can re-write eqn (3.21) as

$$R_c = a^*F_c + bW. \qquad (3.22)$$

The term $a(dW/dt)$ (or a^*F_c) is known as the 'growth respiration' and bW is known as the 'maintenance respiration'. Equations (3.21) and (3.22) imply that the canopy 'dark' respiration rate will increase in proportion to the standing crop mass. However, the data shown in Table 3.2 indicate that although the leaf area index of a 'closed crop' of chrysanthemum plants may increase four-fold, with a similar increase in standing crop mass, there is no perceptible effect on the canopy 'dark' respiration rate.

With rare exceptions the processes of 'dark' respiration, or more correctly oxidative phosphorylation, are tightly coupled with the metabolic process occurring in plant tissues. It is not unreasonable to suppose that the metabolic activities of these tissues will largely depend on the absolute level of carbon assimilation, photosynthesis, by the plant. It is apparent from Fig. 3.6 that once a 'closed crop' has attained a leaf area index of between 2 and 3, further leaf production will only increase the rate of canopy photosynthesis (canopy or crop metabolic activity?) by up to about 30%. It has been argued that the term bW might be replaced by $b*N$, where N is the plant/crop nitrogen or protein mass per unit ground area. We might expect N to remain reasonably constant as the crop approaches its maximum potential photosynthetic performance.

The models for canopy net photosynthesis described in the previous sections have used the single term R_c to describe the rate of canopy 'dark' respiration. Where the models have been used to simulate canopy net photosynthesis or 'fitted' to canopy net photosynthesis data an experimentally measured value of R_c has been used as a constant dependent variable. Whereas it is not difficult to explicitly account for the contribution of leaf respiration, the canopy respiration rate includes a component due to the non-photosynthetic tissues such as roots and fruits. The leaf component of the canopy 'dark' respiration rate, $(R_c)_L$, can be written as

$$(R_c)_L = \int_0^L (R_d)_0 \exp(-KL)\, dL = \frac{(R_d)_0}{K}(1 - \exp(-KL)), \qquad (3.23)$$

where $(R_d)_0$ is the 'dark' respiration rate of an unshaded leaf at the top of the canopy. Equation (3.23) is based on the assumption that the leaf 'dark' respiration rate is a function of the light environment of the leaf when it is *in situ*

TABLE 3.9. Measured and predicted values of the leaf component of the canopy 'dark' respiration rate for a 'closed' crop of tomato plants. Respiration rates were measured and predicted for the crop following progressive defoliation from the ground upwards. The canopy extinction coefficient, K, and leaf area index, L, of the complete canopy, and following the two defoliation treatments, are also shown.

	K	L	Measured $(R_c)_L/$ 10^{-3} g m^{-2} s^{-1}	Predicted $(R_c)_L/$ 10^{-3} g m^{-2} s^{-1}
Complete canopy	0.52	8.6	0.12	0.14
After first defoliation	0.57	5.2	0.11	0.12
After second defoliation	0.63	2.0	0.07	0.08

in the canopy (cf. eqn (3.15)). Measured and predicted values for the leaf component of the 'dark' respiration rate of a 'closed crop' of tomato plants are compared in Table 3.9. Values of the leaf component were obtained for three different leaf layers, and these were separately predicted using the measured values of K and L, shown in the table, and a value $(R_d)_0$ of 0.072×10^{-3} g m^{-2} s^{-1} derived from measurements on single leaves of tomato plants experiencing similar light regimens to those of upper unshaded leaves in the 'closed crop' canopy.

3.6. THE RADIATION ENVIRONMENT WITHIN ISOLATED PLANT STANDS

The models for canopy net photosynthesis described so far have been for 'closed crops', that is, we have assumed in their derivation that the downward light flux density shows no spatial variation in the horizontal plane. However, many horticultural crops are grown as blocks or rows of plants, and the penetration of light between adjacent blocks or rows can lead to considerable spatial inhomogeneity in the downward light flux density. This inhomogeneity will clearly affect the photosynthetic performances of these crops, and we need to develop some mathematical understanding of this variability.

Let us suppose that the plant canopy consists of a large number of small leaves which are randomly distributed with a constant probability throughout the volume of the canopy. Then if $A(x, y, z)$ is the leaf area density (that is leaf area per unit volume of space),

$$A(x, y, z) = A = \text{constant}. \tag{3.24}$$

Now if we consider two points P and Q, and let I_P^* and I_Q^* be the light *intensities* (W m^{-2} sr^{-1}) in the direction QP at the points P and Q respectively, we can write

$$I_P^* = I_Q^* \exp(-kL), \tag{3.25}$$

where L is the leaf area index traversed along the path QP projected onto a plane normal to QP and k is an extinction coefficient. The point $P(x_P, y_P, z_P)$ may lie inside or outside the canopy, and we are interested in the light intensity at P, denoted by I_P^*, from the direction through the point $Q(x_Q, y_Q, z_Q)$ at the surface of the plant. Since we are concerned only with the downward light flux density through a horizontal plane at P, if P lies within the plant $z_Q > z_P$. If P lies outside the plant canopy the path QP intersects the canopy surface at a second point $N(x_N, y_N, z_N)$ where $z_N, z_Q \geqslant z_P$. Now, if P lies within the canopy the pathlength over which the light is attenuated is QP, whereas if P lies outside the canopy it is NQ. In either case we can denote the pathlength by s. The leaf area index L^* traversed along the pathlength s is obtained by integrating the leaf area density function $A(x, y, z)$ along the path, that is

$$L^* = \int_0^s A(x, y, z)\,ds, \qquad (3.26)$$

but since $A(x, y, z)$ is taken to be constant (eqn (3.24)), eqn (3.26) reduces to

$$L^* = As. \qquad (3.27)$$

For randomly inclined and oriented leaves the projected leaf area index, L, is given by

$$L = \tfrac{1}{2}L^* = \tfrac{1}{2}As, \qquad (3.28)$$

and substitution into eqn (3.25) then leads to

$$I_P^* = I_Q^* \exp(-\tfrac{1}{2}kAs). \qquad (3.29)$$

We can then write that the downward light flux density through a horizontal plane at the point P, I_P, is

$$I_P = \int_0^{2\pi} \int_0^{\frac{1}{2}\pi} B \exp(-\tfrac{1}{2}kAs) \cos\theta \sin\theta\,d\theta\,d\phi, \qquad (3.30)$$

where $B(\theta, \phi)$ is a function describing the luminosity of the sky (W m^{-2} sr^{-1}) along the direction specified by (θ, ϕ). Simply, eqn (3.30) formally represents the contributions to the downward light flux density at P of the separate *intensities* from all directions in the solid angle (θ, ϕ). The main problem now is that the pathlength s depends on the angles θ, ϕ, the co-ordinates of the point P and the

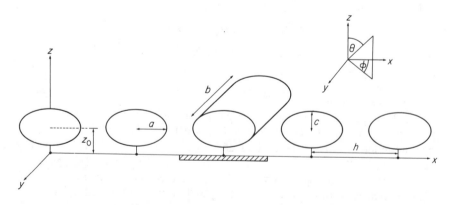

FIG. 3.8. A diagrammatic representation of a cross-section, in the xy plane, of a model of an apple orchard. The orchard is assumed to consist of five ellipsoids, with semi-axes a, b, c in the x, y and z directions. In order to simulate rows b is given a much larger value than a or c. The simulated transect is shown under the centre row.

shape of the canopy. For a standard overcast sky we can write B as a function of θ such that

$$B = \tfrac{1}{3}B_0(1 + 2 \cos \theta), \tag{3.31a}$$

which assumes that the luminosity of the sky at the zenith is three times its luminosity at the horizon. In the absence of any obstructions, integration of eqn (3.31a) gives the downward light flux density on a horizontal surface as

$$I_0 = 7\pi B_0/9 \tag{3.31b}$$

and, B in eqn (3.30) can be replaced by

$$B = (1 + 2 \cos \theta)(3I_0/7\pi). \tag{3.32}$$

There may be an additional component to the downward light flux density at P due to direct sunlight. If I_s is the unobstructed light flux density through a plane perpendicular to the sun's direction, this component to I_P will be

$$I_s \cos \theta_s \exp(-\tfrac{1}{2}kAs), \tag{3.33}$$

where θ_s is the angle of the sun's elevation.

The remaining problem is that of defining the physical shape of the plant canopy. For example, we could represent an orchard as five parallel rows of trees with ellipsoidal canopies. A cross-section of a model orchard canopy in the xz plane is shown in Fig. 3.8. The physical shape of the canopy is then described by the equations

$$\frac{(x - nh)^2}{a^2} + \frac{y^2}{b^2} + \frac{(z - z_0)^2}{c^2} = 1, \qquad n = 0, 1, 2, 3, 4, \tag{3.34}$$

where h is the row width, a, c and b are the three semi-axes of the ellipsoid whose centre is at the point $x = 0$, $y = 0$ and $z = z_0$ and n is the number of rows of trees in the orchard. If P is the point at which the proportion of radiation transmitted is required and Q is a point at the surface of the canopy, then the co-ordinates of P and Q must satisfy the equations

$$\frac{x_Q - x_P}{\sin \theta \cos \phi} = \frac{y_Q - y_P}{\sin \theta \sin \phi} = \frac{z_Q - z_P}{\cos \theta}. \tag{3.35}$$

If the desired transmission profile is at ground level the total pathlength of light within the canopy, s, along the direction QP is given by

$$s = (z_Q - z_N)/\cos \theta, \tag{3.36}$$

where N is the point where the line QP intersects the lower surface of the canopy. For any pairs of values (θ, ϕ) the values of z_Q and z_N can be obtained from eqns (3.34) and (3.35) by co-ordinate geometry. The values of s then

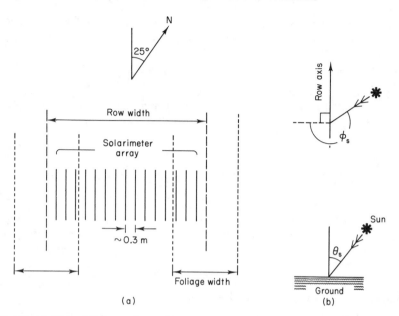

FIG. 3.9. (a) Plan of the experimental arrangement to obtain mean transmission between two rows. (b) Definition of sun position (θ_s, ϕ_s) measured with respect to the row direction.

obtained for pairs of values (θ, ϕ) using eqn (3.36) can be used to integrate eqn (3.30) numerically. These somewhat cumbersome calculations are conveniently carried out by a computer. The proportion of light transmitted to the ground at a point beneath the canopy is then simply the ratio of $I_P : I_0$, where both I_P and I_0 may include a component due to direct sunlight.

Although the mathematics are daunting, the approach is quite readily applied to real problems. Radiation transmission profiles have been obtained at ground level across a row in a hedgerow apple orchard. The experimental arrangement used to obtain the profile is shown in Fig. 3.9. The orchard consisted of 8-year-

TABLE 3.10. The proportions of direct beam (I_B) and diffuse (I_D) radiation to total radiation (I_T), sun zenithal angle (θ_s) and azimuthal angle with respect to the normal to the row axis (ϕ_s) on three occasions that radiation transmission profiles were measured in the hedgerow apple orchard.

I_B/I_T	I_D/I_T	$\theta_s/°$	$\phi_s/°$
0	1.00	—	—
0.63	0.37	50	300
0.56	0.44	35	248

old trees in a hedgerow system. The rows were about 60 m long, oriented 25° west of north. In the transverse direction the orchard was about 300 m wide. The rows were 4.5 m apart, and at the time of the experiment the trees formed an almost continuous hedge about 2 m high. There was no foliage below 0.4 m above ground level, the mean foliage width was 1.8 m and the leaf area density, F, 1.8 m^{-1}. The radiation transmission profiles were measured on three occasions: on an overcast day when there was no direct sunlight and twice on a day when there was a considerable component of direct sunlight. The proportions of diffuse and direct beam radiation in the incoming radiation and the sun angles are shown in Table 3.10. The parameters of the ellipsoidal rows of the model orchard (see Fig. 3.8) were obtained directly by measurement of the experimental orchard. The only undetermined parameter, the canopy extinction coefficient

TABLE 3.11. Parameter values used to simulate the radiation transmission profiles in the hedgerow apple orchard.

Semi-axes of the ellipsoid	
a	0.9 m
b	50.0 m
c	0.8 m
z_0	1.2 m
Canopy extinction coefficient, K	1.0
Leaf area density, A	1.8 m^{-1}

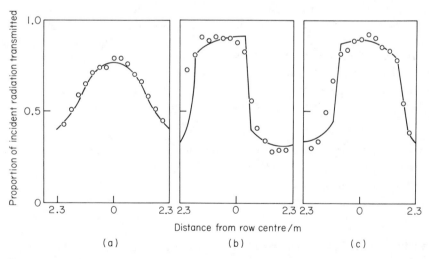

FIG. 3.10. Simulated (continuous line) and measured (circles) transmission of radiation to ground level below the hedgerow orchard; (a) when there was no direct beam radiation, and (b), (c) on two occasions when there was a strong direct beam component. The details are given in Tables 3.10 and 3.11.

k, was assumed to be unity. The parameter values used are shown in Table 3.11. The simulated transmission profiles for the transect in the x direction for values x_P from $3h/2$ to $5h/2$ at $y_P = 0$, $z_P = 0$ with diffuse and direct beam components corresponding to those given in Table 3.10 are compared with the experimentally measured profiles in Fig. 3.10. The simulated and observed profiles are in good agreement.

On the basis of these data a more detailed specification of leaf positioning within the canopy appears unnecessary, although on these apple trees there was obvious clumping of leaves into clusters on old wood and other leaves on to long extension shoots. The most significant clumping in the hedgerow orchard is, however, into rows, and is adequately described by the model. For these data further specification of non-randomness, or an explicit treatment of radiation scattering, does not appear to be justified.

3.7. PHOTOSYNTHESIS BY ISOLATED PLANT STANDS

It has been shown in the previous section how we might develop simple, semi-empirical models to describe the light environment within isolated plant stands. If we were to divide the plant canopy into a number of small elements we could use this approach to calculate the downward light flux density at the mid-point of each element. If we also knew the photosynthetic characteristics of the leaves within each element it would be no great problem to calculate the photosynthetic rate of each element for any particular downward light flux density above the canopy. The photosynthetic rate of the whole canopy can then be quite simply obtained by summing the contributions of each of the elements. Let us suppose that the canopy is divided into i elements each of volume v, and that the downward light flux density at the centre of each element is I_i. We can write the photosynthetic rate of any leaf within the ith volume, F_i, as

$$F_i = \text{function}\,(I_i), \tag{3.37}$$

where function (I_i) might be any one of the equations for leaf photosynthesis described in the previous sections. The rate of canopy net photosynthesis per unit of ground area, F_c, is then given by

$$F_c = \sum_{0}^{i} F_i A v/G, \tag{3.38}$$

where A is the leaf area density and G is the ground area covered by the stand of plants.

The data given in Table 3.5 suggest, at least for tomato leaves, that the individual leaf photosynthetic characteristics may differ for leaves from different positions within the crop canopy. As a first approximation, however, we could

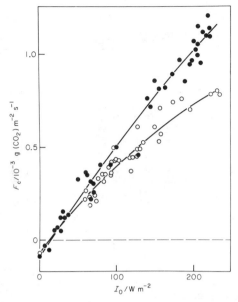

FIG. 3.11. Measured canopy net photosynthetic rate versus light response curves for a 'closed crop' of chrysanthemum plants at two ambient carbon dioxide concentrations (○, 0.74 g $(CO_2)m^{-3}$; ●, 2.20 g $(CO_2)m^{-3}$). The solid lines were obtained by 'fitting' eqn (3.9) to the combined data sets.

use eqns (3.37) and (3.38) by assuming that the photosynthetic characteristics of all leaves within the canopy were the same.

The 'average' leaf photosynthetic parameters for a 'closed' crop of chrysanthemum plants have been estimated by 'fitting' eqn (3.9) to canopy photosynthesis data. Chrysanthemum plants were grown as a 'closed' crop in daylight assimilation chambers, and the canopy net photosynthesis versus light response curves of two ambient carbon dioxide levels were obtained. The data are shown in Fig. 3.11, and the solid lines represent the curves obtained by 'fitting' eqn (3.9) to them. Values of the 'average' leaf photosynthetic parameters

TABLE 3.12. Numerical estimates of the 'average' leaf photosynthetic parameters α_m, β, τ and the measured values of L, K and R_c for the closed chrysanthemum crop.

α_m	9.9 (±0.6) × 10^{-6} g (CO_2) J^{-1}
β	2.1 (±2.5) × 10^{-4} g (CO_2) m^{-2} s^{-1}
τ	2.7 (±1.1) × 10^{-3} m s^{-1}
L	4.18
K	0.71
R_c	0.075 × 10^{-3} g m^{-2} s^{-1} at 0.736 g (CO_2) m^{-3}
	0.095 × 10^{-3} g m^{-2} s^{-1} at 2.21 g (CO_2) m^{-3}

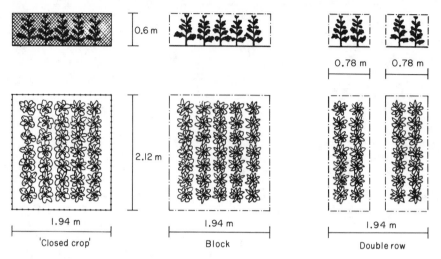

FIG. 3.12. Diagrammatic representation of the canopy arrangements used to simulate a 'closed' crop, an isolated block of plants and a double row crop for canopy photosynthesis measurements in a daylight assimilation chamber.

α_m, β and τ, together with measured values of the canopy characteristics L, K and R_c are given in Table 3.12.

The plants were also arranged as an isolated stand and double row, and their canopy photosynthetic rates recorded. The experimental configuration is shown in Fig. 3.12.

It was assumed that the crop canopies of the isolated stand and double row were described by

$$x = \pm a + nh, \qquad n = 0, 1 \tag{3.39}$$

$$y = \pm b \tag{3.40}$$

$$z = c, \tag{3.41}$$

where a and b are the half widths of the canopy in the horizontal plane, c is the height and h, in the case of the double row canopy, is the distance between row centres. If P is the point at which the downward light flux density within the canopy is required, and Q is a point at the surface of the canopy, then the co-ordinates of P and Q must satisfy the eqns (3.35), and the pathlength of light, s, is given by

$$s = (z_Q - z_P)/\cos \theta. \tag{3.42}$$

For any pairs of values (θ, ϕ) the values of z_Q and z_P can be obtained from eqns (3.35), and thence s from eqn (3.42). If we use eqns (3.30)–(3.32) to describe I_P, the downward light flux density at point P, values of s for particular

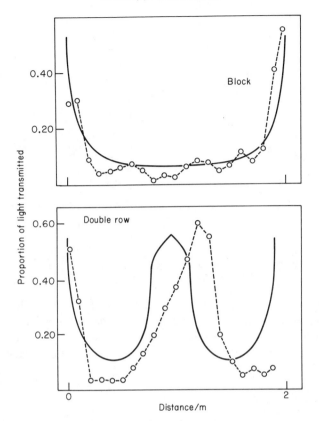

FIG. 3.13. Light transmission at the bases of both an isolated block and double row crop of chrysanthemum plants. The ○ denotes the measured values and the solid lines the simulated transmission profiles. The 'skewness' of the measured transects could be attributed to a directional component in the incident light not taken into account in the simulations.

values of (θ, ϕ) can be used to solve them for I_P. Measured and simulated transmission profiles at the bases of the isolated block and double row canopies of chrysanthemum plants are shown in Fig. 3.13.

Using the single leaf photosynthesis equation

$$F_i = \alpha_m I_i (\tau C - \beta)/(\alpha_m I_i + \tau C) - (R_d)_i \tag{3.43}$$

the photosynthesis vs. light response curves for canopy net photosynthesis at 0.74 g (CO_2) m^{-3} ambient carbon dioxide concentration were simulated, for the 'average' parameter values shown in Table 3.12, for the isolated block and double row canopies. The simulations are compared with the experimentally observed responses in Fig. 3.14.

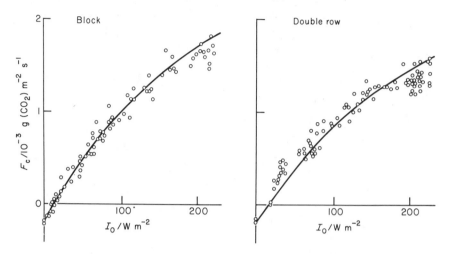

FIG. 3.14. Simulated (solid lines) and measured (O) canopy photosynthesis vs. incident light response curves for an isolated block and double row of chrysanthemum plants.

At the level of resolution available for canopy net photosynthesis measurements, the semi-empirical light interception model described in the previous section enables a reasonable prediction to be made of the performance of isolated plant stands from a knowledge of their leaf photosynthetic performances.

3.8. DISCUSSION

Separate mathematical descriptions of (a) the instantaneous response of the rate of leaf net photosynthesis to changes in the leaf's environment and (b) the different environments of the individual leaves within the crop canopy, can be combined to provide simple, but useful, quantitative mathematical analyses and predictors of the canopy's photosynthetic behaviour. For the 'closed' crop, where it can be assumed that the major source of intra-canopy environmental variation lies in a varying downward light flux density in the vertical plane, the mathematical behaviour of canopy photosynthesis can be described in terms of the photosynthetic behaviour of a well-defined leaf, an upper unshaded leaf (cf. eqn (3.18)). The markedly non-linear response of the canopy photosynthetic rate to changing incident light flux densities can be simply and usefully 'linearized', so that simple linear regression methods can be used to analyse a variety of 'closed' crop data (see Section 3.4). The analysis also provides a simple description of the way in which canopy architecture can modify the canopy's photosynthetic performance. For isolated stands of plants intra-canopy variation in downward light flux density in both the horizontal and vertical planes needs to be taken into account. Computer techniques allow this situation to be dealt with quite readily.

The mathematical description of a process such as canopy photosynthesis enables us to examine the theoretical interaction(s) of the different components of the process. For example, eqn (3.18) predicts that at light saturation

$$F_{c(MAX)} \simeq \frac{\tau_0 C}{K} \left[1 - \exp(-KL)\right]. \tag{3.44}$$

Expansion of the right hand side of eqn (3.44) leads to

$$F_{c(MAX)} \simeq \tau_0 CL \left(1 - \frac{KL}{2!} + \frac{KL^2}{3!} - \ldots \right), \tag{3.45}$$

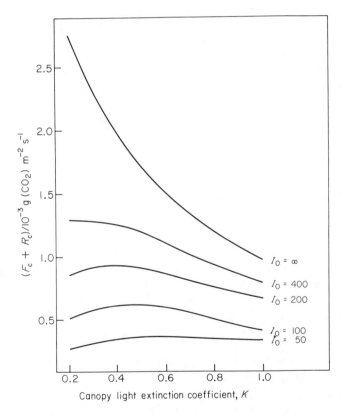

FIG. 3.15. The simulated effects of different canopy extinction coefficients on the photosynthetic rates of 'closed' crops at different incident light levels. The simulations were derived from eqn (3.18) with $\alpha = 10 \times 10^{-6}$ g J^{-1}, $\tau_0 C = 1.0 \times 10^{-3}$ g m^{-2} s^{-1}, $L = 4$ and $m = 0.10$.

which predicts that $F_{c(MAX)}$ will decrease with increasing canopy extinction coefficient. That is, the smaller the canopy extinction coefficient, or more erectophile the leaf habit, the greater the rate of light-saturated canopy photosynthesis at any particular leaf area index. However, at other than saturating incident light flux densities this will not be so. The dependence of F_c on K at different incident light flux densities for a 'closed' crop at one leaf area index is shown in Fig. 3.15. The simulations, derived from eqn (3.18), demonstrate that there may be an optimal canopy extinction coefficient which changes with the incident light flux density. Equation (3.45) also demonstrates an interaction between K and L. The optimal value of K for a particular crop will therefore depend on both the leaf area index of the crop and the 'average' incident light flux density it experiences in the field. In general, the lower the light the more productive crop will have planophile leaves. Equally, we can use eqn (3.44) to show that the instantaneous rate of light-saturated canopy photosynthesis will be linearly dependent on the numerical value of τ_0. A 'closed' crop composed of genotypes with high rates of light-saturated photosynthesis will, according to eqn (3.44), have a high rate of canopy light-saturated photosynthesis. Most usefully, the derivation of eqn (3.18), and its similarity to the single leaf response, allows us to suppose that the rate of net photosynthesis by a 'closed' crop canopy will respond to changes in the aerial environment, and other factors such as increasing leaf water deficits, in an analogous manner to the rate of net photosynthesis by a single leaf.

3.9. MAIN SYMBOLS

Symbol	Meaning	Unit
C	ambient carbon dioxide concentration	$g\ m^{-3}$
I_0	downward light flux density above the crop	$W\ m^{-2}$ or $J\ m^{-2}\ s^{-1}$
F	leaf net photosynthetic rate per unit leaf area	$g\ m^{-2}\ s^{-1}$
R_d	net rate of 'dark' respiration per unit leaf area	$g\ m^{-2}\ s^{-1}$
F_c	net canopy photosynthetic rate per unit ground area	$g\ m^{-2}\ s^{-1}$
R_c	net canopy 'dark' respiration rate per unit ground area	$g\ m^{-2}\ s^{-1}$
α_m	maximum leaf light utilization efficiency	$g\ J^{-1}$
α	actual leaf light utilization efficiency	$g\ J^{-1}$
α_c	canopy light utilization coefficient	$g\ J^{-1}$
τ	overall leaf carboxylation constant	$m\ s^{-1}$
β	leaf photorespiration constant	$g\ m^{-2}\ s^{-1}$
τ_0	overall carboxylation constant of an unshaded leaf	$m\ s^{-1}$
K	canopy light extinction coefficient	

L leaf area index
m leaf light transmission coefficient
s_A specific leaf area $m^2\ g^{-1}$
A leaf area density $m^2\ m^{-3}$ or m^{-1}
θ_s sun's zenithal angle
ϕ_s sun's azimuthal angle

3.10. SUGGESTED FURTHER READING

Section 3.2

Charles-Edwards, D. A. and Acock, B. (1977). *Ann. Bot.* **41**, 49–58.
Monsi, M. and Saeki, T. (1953). *Jap. J. Bot.* **14**, 22–52.
Saeki, T. (1960). *Bot. Mag. Tokyo* **73**, 155–163.
Thornley, J. H. M. (1976). *In* 'Mathematical Models in Plant Physiology', Ch. 3. Academic Press, London and New York.

Section 3.3

Acock, B., Charles-Edwards, D. A. and Hand, D. W. (1976). *J. exp. Bot.* **27**, 933–941.
Acock, B., Charles-Edwards, D. A. and Sawyer, S. (1979). *Ann. Bot.* **44**, 289–300.
Acock, B., Charles-Edwards, D. A., Fitter, D. J., Hand, D. W., Ludwig, L. J., Warren Wilson, J. and Withers, A. C. (1978). *J. exp. Bot.* **29**, 815–827.
Charles-Edwards, D. A. and Acock, B. (1977). *Ann. Bot.* **41**, 49–58.

Section 3.4

Acock, B., Charles-Edwards, D. A., Fitter, D. J., Hand, D. W., Ludwig, L. J., Warren Wilson, J. and Withers, A. C. (1978). *J. exp. Bot.* **29**, 815–827.
Charles-Edwards, D. A. (1979). *In* 'Photosynthesis and Plant Development' (R. Marcelle, H. Clijsters and M. Van Pouke, eds). Junk, The Hague.
Hozumi, K., Kinta, M. and Nishioka, M. (1972). *Photosynthetica* **6**, 158–168.
Kira, T. (1975). *In* 'Biosynthesis and Productivity in Different Environments' (J.P. Cooper, ed.). Cambridge Univ. Press, London.
Ludlow, M. M. and Charles-Edwards, D. A. (1980). *Aust. J. Agric. Res.* **31**, 673–692.

Section 3.5

Barnes, A. and Hole, C. C. (1978). *Ann. Bot.* **42**, 1217–1221.
Charles-Edwards, D. A. and Acock, B. (1977). *Ann. Bot.* **41**, 49–58.
Thornley, J. H. M. (1976). *In* 'Mathematical Models in Plant Physiology', Ch. 6. Academic Press, London and New York.
Thornley, J. H. M. (1977). *Ann. Bot.* **41**, 1191–1203.

Section 3.6

Charles-Edwards, D. A. and Thornley, J. H. M. (1973). *Ann. Bot.* **37**, 919–928.
Charles-Edwards, D. A. and Thorpe, M. R. (1976). *Ann. Bot.* **40**, 603–613.

Section 3.7

Acock, B., Charles-Edwards, D. A., Fitter, D. J., Hand, D. W. and Warren Wilson, J. (1978). *Ann. appl. Biol.* **90**, 255–263.

Section 3.8

Acock, B., Hand, D. W., Thornley, J. H. M. and Warren Wilson, J. (1975). *Ann. Bot.* **40**, 1293–1307.

4. Photosynthesis and Productivity

4.1. INTRODUCTION

A mathematical model of leaf photosynthesis was developed in Chapter 2. The model is based on our 'guestimate' of the global kinetic behaviour of the underlying biochemical and physical processes of leaf photosynthesis. In Chapter 3 the leaf photosynthesis model was simplified and used to develop mathematical models of canopy photosynthesis. These models describe canopy photosynthesis in terms of the assumed photosynthetic characteristics of the component leaves of the canopy. It was shown that these models provide useful qualitative descriptions of the markedly non-linear responses of the canopy net photosynthetic rate to changes in the independent variables, such as the diurnal variation in light flux density incident at the top of the canopy. The numerical values of the leaf photosynthetic parameters giving rise to the 'best fit' of the canopy model to experimental data are not dissimilar, both in their numerical values and in their responses to changing growth regimens, to the 'equivalent' parameters estimated directly from experimental data for photosynthesis by single leaves. We therefore have some reason to believe that our mathematics have successfully integrated our understanding of the photosynthetic behaviour of single leaves into an understanding of the photosynthetic behaviour of a community of plants. It is logical now to attempt to integrate this understanding into a mathematical model for crop growth and productivity.

It was pointed out at the start of this book (see Section 1.3) that photosynthesis is only one of a number of processes by which plants acquire material from their environment. In many agricultural situations it is the interaction between these primary assimilating processes which will determine plant/crop growth. This interaction can be demonstrated theoretically for seedlings growing experimentally in the 'steady-state', and this is done in Section 4.2. Only when all other essential nutrients are supplied to the seedling in non-limiting amounts can carbon acquisition, or photosynthesis, be said to limit growth. This situation can be attained experimentally. In these circumstances the canopy photosynthesis models developed in Chapter 3 could be used to predict the gain in *total* dry weight by a 'closed crop'. The derivation of a simple model and comparison of its predictions with experimental observation are described in Section 4.3.

The root/shoot interactions described in Section 4.2 can be used as a semi-empirical basis for a more complete analysis of crop growth in the field. A mathematical model for crop growth and productivity is developed in Section 4.4. The model extends the analysis presented in Section 4.3 to predict not only *total* dry matter production but also the production of the separate crop components; leaves, stems, roots etc. Prediction of the daily gain in carbon, the daily photosynthetic integral, remains an important component of this new model. A simple mathematical description of this integral is obtained in Section 4.5. This mathematical description explicitly allows for intra-canopy variation in leaf photosynthetic characteristics. It is mathematically analogous to, and comparable with, the mathematical description of the *instantaneous* rate of canopy photosynthesis from which the integral was derived.

The crop growth model can be used as both a predictive and analytical tool. In Section 4.6 it is 'fitted' to field data on crop growth. These data illustrate the physiological and agronomic implications of this type of growth analysis. In the final section, Section 4.7, the relative merits of this type of growth analysis are discussed in the context of the more traditional analysis of crop growth data. The analysis is also used to examine the potential effects of genetic variation in some leaf photosynthetic characteristics on crop growth and productivity.

4.2. ROOT/SHOOT INTERACTIONS

Let us imagine that a plant acquires an amount of dry matter ΔW during the period of time Δt. Let us also suppose that a fraction f_M of that dry matter consists of the element M. We can write, without any approximation or assumption

$$f_M = \Delta M / \Delta W \tag{4.1}$$

and

$$\Delta W / \Delta t = (1/f_M) \, \Delta M / \Delta t. \tag{4.2}$$

If the plant is in the vegetative state its total weight will be the sum of its shoot weight, T, and its root weight, R, so that

$$W = T + R. \tag{4.3}$$

If the element M is carbon we can confidently assume that it is taken up, by photosynthesis, by the aerial parts of the plant alone, and can write

$$\sigma_C = (1/T) \, \Delta C / \Delta t, \tag{4.4}$$

where $\Delta C/\Delta t$ is the rate of carbon uptake by the plant and σ_C the specific shoot activity with respect to carbon (that is the rate of uptake of carbon per unit mass of shoot). Similarly, if the element M is nitrogen, we can write

$$\sigma_N = (1/R)\,\Delta N/\Delta t, \tag{4.5}$$

where $\Delta N/\Delta t$ is the rate of nitrogen uptake by the plant and σ_N the specific root activity with respect to nitrogen. Equations (4.2), (4.4) and (4.5) then give

$$\Delta W/\Delta t = T\sigma_C/f_C = R\sigma_N/f_N. \tag{4.6}$$

Provided σ_C and σ_N are independent of T and R we can write

$$\Delta R = \Delta T\,(f_N\sigma_C/f_C\sigma_N). \tag{4.7}$$

Note that this can hold only with spaced seedlings when there is no significant leaf shading or competition between the roots. Now eqn (4.3) gives rise to

$$\Delta W = \Delta T + \Delta R, \tag{4.8}$$

and from eqn (4.7),

$$\Delta W = \Delta T\,(f_N\sigma_C + f_C\sigma_N)/f_C\sigma_N. \tag{4.9}$$

Substituting for ΔW in eqn (4.6) then leads to

$$\Delta T/\Delta t = T\sigma_C\sigma_N/(f_N\sigma_C + f_C\sigma_N). \tag{4.10}$$

We can show, using simple algebra, that

$$(1/T)\,\Delta T/\Delta t = (1/R)\,\Delta R/\Delta t = (1/W)\,\Delta W/\Delta t = \mu, \tag{4.11}$$

where

$$\mu = \sigma_N\sigma_C/(f_N\sigma_C + f_C\sigma_N). \tag{4.12}$$

Equation (4.12) can be re-written in the form

$$\mu = \left(\frac{f_N}{\sigma_N} + \frac{f_C}{\sigma_C}\right)^{-1}. \tag{4.13}$$

If we consider x different elements assimilated by the roots and y different elements assimilated by the shoots, eqn (4.13) can be generalized to

$$\mu = \frac{xR + yT}{W}\left(\sum_x \frac{f_x}{\sigma_x} + \sum_y \frac{f_y}{\sigma_y}\right)^{-1} \tag{4.14}$$

Equation (4.14) will still hold for spaced seedlings growing in the reproductive state when we can write W as

$$W = T + R + F \tag{4.15}$$

where F is the dry weight of the reproductive parts.

For spaced, vegetative seedlings growing under a constant incident light flux density I during a 'daylength' of Z seconds the rate of carbon assimilation averaged over the entire 24 h day/night period can be written as

$$\frac{\Delta C}{\Delta t} = \frac{0.273\,YLZ}{86\,400}\,\frac{\alpha I \tau C}{\alpha I + \tau C}. \tag{4.16}$$

The factor 0.273 converts g (CO_2) to g (carbon), (= 12/44), Y is a yield factor accounting for carbon dioxide lost through 'dark' respiration, L is the leaf area of the seedling, 86 400 is the number of seconds in 24 h and the term $\alpha I \tau C/(\alpha I + \tau C)$ describes the instantaneous rate of leaf photosynthesis. The symbol C denotes the ambient carbon dioxide concentration, α the light utilization efficiency and τ the overall leaf carboxylation conductance. Combining eqns (4.4) and (4.16) we can write

$$\mu^{-1} = \frac{f_N}{\sigma_N} + \frac{86\,400 f_C\,T}{0.273\,YLZ}\left(\frac{1}{\alpha I} + \frac{1}{\tau C}\right). \tag{4.17}$$

If we further assume that the empirical formula of plant dry matter is (CH_2O), we can calculate f_C as 0.4 g (carbon) per g (plant dry weight), and for a value of $Y = 0.6$ (that is assuming that 40% of the carbon dioxide assimilated by photosynthesis during the light period is respired during the *whole* 24 h period), we can reduce eqn (4.17) to

$$\mu^{-1} = \frac{f_N}{\sigma_N} + 2.11 \times 10^5\,\frac{T}{LZ}\left(\frac{1}{\alpha I} + \frac{1}{\tau C}\right). \tag{4.18}$$

Wheat seedlings were grown in spaced pots in controlled environment rooms at different light levels. The seedlings were provided with 'non-limiting' nutrients by standing the pots in saucers containing nutrient solution. The specific growth rates, leaf area to shoot weight ratios and growth light levels were recorded, and are shown in Table 4.1. Provided the term $1/\tau C$ is far less than $1/\alpha I$, eqn (4.18) predicts a linear relationship between $1/\mu$ and T/LI. The linear regression for the data given in Table 4.1 is shown in Fig. 4.1. The slope of the regression

TABLE 4.1. Specific growth rates, μ, and leaf area to shoot weight ratios, L/T, for wheat seedlings grown in controlled environment rooms with a 16 h light period, but at different mean light flux densities, I.

$I/\mathrm{J\ m^{-2}\ s^{-1}}$	$\mu/\mathrm{s^{-1}}$	$(L/T)/\mathrm{m^2\ g^{-1}}$
8	7.1×10^{-7}	3.3×10^{-2}
16	13.9×10^{-7}	2.6×10^{-2}
30	17.3×10^{-7}	2.3×10^{-2}
53	26.5×10^{-7}	1.8×10^{-2}
85	28.4×10^{-7}	1.5×10^{-2}

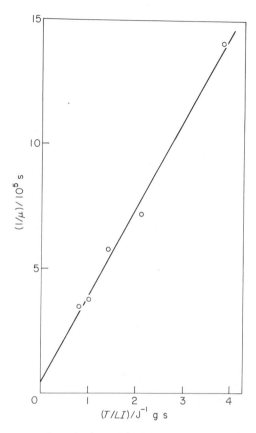

FIG. 4.1. The linear regression of $1/\mu$ on T/LI for the data shown in Table 4.1, and predicted by eqn (4.18).

$(= 2.11 \times 10^5/\alpha Z)$ is 3.5×10^5 J g^{-1}. It is a simple matter to calculate that for these seedlings, grown during a 16 h light period, the leaf light utilization efficiency α is 10.5×10^{-6} g J^{-1}, a value similar to those estimated by analysis of leaf and canopy photosynthesis data for other C$_3$ plants. The intercept obtained from the regression $(= f_N/\sigma_N)$ has a value of 4.5×10^4 s. If we assume that about 5% of the plant dry matter is nitrogen (that is $f_N = 0.05$) we can estimate an approximate value for the specific root activity of the seedlings, σ_N, as 1.1×10^{-6} s^{-1}. This value can be compared with the approximate values of the specific shoot activities of the seedlings which range from 0.2 to 1.2×10^{-2} s^{-1}. These figures confirm that the seedlings were grown with non-limiting nutrient supply.

Although σ_N is far less than σ_C for these wheat seedlings it is likely that they will not differ greatly in many common agricultural situations. Equation (4.12) predicts that the specific growth rate of seedlings growing under steady-state conditions will be hyperbolically dependent on both σ_C and σ_N. The role of photosynthesis as a determinant of the plant's growth rate will often, therefore, be modified by the specific activity of the roots with respect to some other essential nutrient.

4.3. TOTAL DRY MATTER PRODUCTION–A SIMPLE MODEL

We can write, using eqns (4.1) and (4.2) that

$$\Delta W/\Delta t = (1/f_C)\Delta C/\Delta t, \qquad (4.19)$$

where $\Delta C/\Delta t$ denotes the rate of carbon uptake by the plant/crop and f_C is the fraction of carbon in the increment of total dry matter ΔW. If we assume that the period of time Δt corresponds to the 24 h day/night period we can re-write eqn (4.19) as

$$\Delta W/\Delta t = 0.273\,(\nabla_F - \nabla_R)/f_C, \qquad (4.20)$$

where ∇_F is the daily net photosynthetic integral, 0.273 converts g (carbon dioxide) to g (carbon) and ∇_R is the daily respiratory integral, accounting for loss of carbon dioxide through 'dark' respiration during the 24 h day/night period. We can also write that at the end of the ith time period

$$W_i = W_{(i-1)} + \Delta W_i. \qquad (4.21)$$

Equations (4.20) and (4.21) constitute a simple model for predicting the day by day changes in plant/crop total dry matter. The problem is reduced to one of predicting ∇_F and ∇_R, the daily photosynthetic and respiratory integrals.

If we use the model for canopy photosynthesis that assumes uniform leaf photosynthetic characteristics throughout the volume of the canopy we can write, for a 'closed' crop

$$\nabla_F = \int_0^Z \frac{\tau C - \beta}{K} \ln\left[\frac{\alpha_m I_0 + (1-m)\tau C}{\alpha_m I_0 \exp(-KL) + (1-m)\bar{\tau}C}\right]dt, \qquad (4.22)$$

where Z is the daylength, C and I_0 are the ambient carbon dioxide concentration and downward light flux density at the top of the canopy, α_m, τ and β are the 'average' leaf light utilization efficiency, overall carboxylation conductance and photorespiration constant, K is a canopy extinction coefficient, L the leaf area index of the crop, and m an 'average' leaf transmission coefficient. If we assume that the diurnal variation in downward light flux density at the top of the canopy is described by

$$I_0 = S[1 + \sin(2\pi t/Z + 3\pi/2)]/Z, \qquad 0 \leqslant t \leqslant Z \qquad (4.23)$$

$$I_0 = 0, \qquad t > Z, \tag{4.24}$$

where S is the daily light integral, we can substitute for I_0 in eqn (4.22) and integrate over the period $t = 0$ to $t = Z$ to obtain

$$V_F = \frac{Z\,(\tau C - \beta)}{K} \ln \left[\frac{A + \sqrt{(A^2 - B^2)}}{E + \sqrt{(E^2 - D^2)}}\right], \tag{4.25}$$

where $B = K\alpha_m S/Z$
$A = B + (1 - m)\,\tau C$
$D = B\exp(-KL)$
$E = D + (1 - m)\tau C$.

Equation (4.25) describes the daily photosynthetic integral as a function of the leaf area index, L, of the crop. As the total dry weight of a vegetative crop increases we might expect L to increase. As a first approximation we can write

$$L = aW + b. \tag{4.26}$$

The daily respiratory integral, V_R, can be written as

$$V_R = \int_0^{86\,400} R_C\,dt \tag{4.27}$$

where R_C is the instantaneous rate of dark respiration.

Equations (4.20), (4.21), (4.25), (4.26) and (4.27) together constitute a complete model for total dry matter production by a 'closed' crop. The model has three independent, environmental variables, Z, S and C, three dependent variables, W, L and V_R, and eight parameters, α_m, β, τ, m, K, f_C, a and b. Values of the eight parameters, and the dependent variable V_R, derived from both dry matter analysis and analysis of canopy photosynthesis data, for a 'closed', vegetative chrysanthemum crop are given in Table 4.2. These numerical values can be used to predict the growth of the crop, and the

TABLE 4.2. Numerical estimates of the parameters α_m, β, τ, m, K, f_C, a and b, and the dependent variable V_R for a chrysanthemum crop growing under natural lighting in a daylight assimilation chamber.

α_m	12.0×10^{-6} g (CO_2) J (total radiation)$^{-1}$
β	1.6×10^{-7} g (CO_2) m^{-2} s^{-1}
τ	1.3×10^{-3} m s^{-1}
m	0.009
K	0.56
f_C	0.41
a	11.8×10^{-3} m^2 g^{-1}
b	0.17 m^2 m^{-2}
V_R	19 g (CO_2) d^{-1}

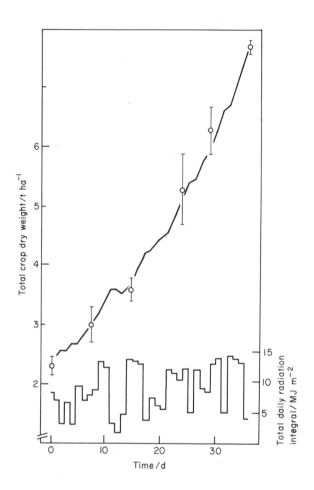

FIG. 4.2. Simulated (solid line) and observed changes in total dry matter of a chrysanthemum crop growing in natural daylight. The daily radiation integrals are also shown in the histogram.

prediction is compared with experimental observation in Fig. 4.2. Note that both the predictions, and the assumed parameter values, are based on total radiation integrals rather than the daily integrals of photosynthetically active light. The model appears to predict the growth of the experimental crop quite successfully.

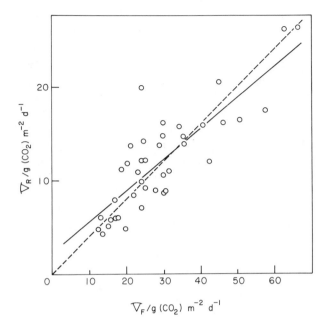

FIG. 4.3. The relationship between ∇_F and ∇_R for a 'closed' chrysanthemum crop growing in natural daylight. The solid line is the fitted linear regression line, and the dashed line is defined by $\nabla_R = 0.4\nabla_F$.

4.4. DAILY PHOTOSYNTHETIC INTEGRALS

It should be apparent that central to any mathematical description of crop growth is the need to calculate crop daily photosynthetic integrals. In the previous section a mathematical description of the daily photosynthetic integral was derived in terms of the 'average' photosynthetic characteristics of the leaves which constitute the crop canopy. As well as being somewhat cumbersome, eqn (4.25) also begs the question 'what is the "average" leaf?' It is attractive to look for a simpler mathematical formulation which describes the daily photosynthetic integral in terms of the photosynthetic character-istics of a 'defined' leaf within the canopy.

In the first instance we can simplify eqn (4.20) by writing

$$\nabla_R = (1 - Y)\,\nabla_F \qquad\qquad (4.28)$$

and hence,

$$\Delta W/\Delta t = 0.273\,Y\nabla_F/f_C \qquad\qquad (4.29)$$

where Y is a 'yield factor' such that the proportion $(1 - Y)$ of ∇_F is the proportion of newly assimilated carbon dioxide respired during the entire 24 h day/night period. Measurements of both ∇_F and ∇_R for a vegetative chrysanthemum crop growing under natural daylight in controlled environment cabinets provide some support for eqn (4.28). Data for the chrysanthemum crop are shown in Fig. 4.3. The scatter in the relationship shown in Fig. 4.3 is to be expected. It is not unreasonable to suppose that carbonaceous materials synthesized during previous light periods provide as great a proportion of the of the substrate for oxidative phosphorylation ('dark' respiration) as materials synthesized during the current light period. It would then follow that during a dull day following a series of bright days, which have led to the 'filling' of metabolite storage pools, we would observe a greater value for ∇_R than might be predicted from the current net photosynthetic integral (eqn (4.28)). Conversely, on a bright day following a series of dull days, when metabolite storage pools have been depleted, we might expect ∇_R to be less than the value predicted by eqn (4.28). Averaged over the growth period of the crop, however, eqn (4.28) may provide a reasonable and simple estimate of ∇_R.

Now eqn (3.23) describes the instantaneous rate of canopy net photosynthesis as

$$F_C + R_C = \frac{\alpha I_0 \tau_0 C[1 - \exp(-KL)]}{K\alpha I_0 + (1 - m)\tau_0 C}, \qquad (4.30)$$

where α and τ_0 are the light utilization efficiency and overall leaf carboxylation conductance of an *upper unshaded leaf* in the 'closed' crop canopy. If we ignore any possible variation in the ambient carbon dioxide concentration, C, and assume that the natural diurnal variation in the downward light flux density at the top of the canopy, I_0, shows a half sine wave response with time, we can write

$$I_0 = (\pi S/2Z)\sin(\pi t/Z), \qquad 0 \leqslant t \leqslant Z, \qquad (4.31)$$

and

$$I_0 = 0, t > Z, \qquad (4.32)$$

where Z is the daylength. Substitution of eqn (4.31) into (4.30) gives

$$F_C + R_C = \frac{\tau_0 C[1 - \exp(-KL)] \sin(\pi t/Z)}{K[\delta + \sin(\pi t/Z)]}, \qquad 0 \leqslant t \leqslant Z, \qquad (4.33)$$

where

$$\delta = 2Z(1 - m)\tau_0 C/K\pi\alpha S. \qquad (4.34)$$

Now,

$$\nabla_F = \int_0^Z (F_C + R_C)\, dt \qquad (4.35)$$

$$= \frac{\tau_0 C[1 - \exp(-KL)]}{K} \int_0^Z \frac{\sin(\pi t/Z)\, dt}{\delta + \sin(\pi t/Z)}. \qquad (4.36)$$

If we replace $\pi t/Z$ by the dummy variable x, then $dt = Z\, dx/\pi$, and we can write

$$\int_0^Z \frac{\sin(\pi t/Z)\, dt}{\delta + \sin(\pi t/Z)} = \frac{Z}{\pi} \int_0^\pi \frac{\sin(x)\, dx}{\delta + \sin(x)} = \frac{2Z}{\pi} \int_0^{\pi/2} \frac{\sin(x)\, dx}{\delta + \sin(x)}$$

$$= \frac{2Z}{\pi} \left(\frac{\pi}{2} - \epsilon \right), \qquad (4.37),$$

where

$$\epsilon = \frac{2\delta}{\sqrt{(\delta^2 - 1)}} \, \mathrm{arctg} \left[\frac{\sqrt{(\delta^2 - 1)}}{\delta + 1} \right], \qquad \delta^2 > 1 \qquad (4.38)$$

$$\epsilon = 1, \qquad \delta^2 = 1 \qquad (4.39)$$

$$\epsilon = \frac{\delta}{\sqrt{(1 - \delta^2)}} \ln \left[\frac{1 + \delta + \sqrt{(1 - \delta^2)}}{1 + \delta - \sqrt{(1 - \delta^2)}} \right], \qquad \delta^2 < 1. \qquad (4.40)$$

Numerical values of ϵ can be calculated for all values of δ from $0 \to \infty$. These values, together with the values of $\pi/2 - \epsilon$ are shown in Table 4.3. A good approximation to $\pi/2 - \epsilon$ over the whole range of δ from $0 \to \infty$ is given by $(\pi/2)/[1 + (\delta\pi/2)]$. Numerical values of this algorithm for different values of δ are also shown in Table 4.3. As a first approximation, therefore, we can write

$$\frac{\pi}{2} - \epsilon = \frac{\pi/2}{1 + (\delta\pi/2)}, \qquad (4.41)$$

and substituting for $\pi/2 - \epsilon$ in eqn (4.37) using eqn (4.41), and thence substituting the integral in eqn (4.36), we can show that as a good approximation

$$\nabla_F = \frac{\alpha S \tau_0 CZ[1 - \exp(-KL)]}{K\alpha S + (1 - m)\tau_0 CZ}. \qquad (4.42)$$

Equation (4.42) is simple, and bears a remarkable similarity to eqn (4.30). If we substitute eqns (4.42) and (4.28) into eqn (4.20) we have

$$\frac{\Delta W}{\Delta t} = \frac{0.273\, Y\alpha S \tau_0 CZ[1 - \exp(-KL)]}{f_C[K\alpha S + (1 - m)\tau_0 CZ]}. \qquad (4.43)$$

TABLE 4.3. Values of ϵ and $\pi/2 - \epsilon$ calculated exactly for different estimates of the parameter δ. Also shown are the comparable values of the algorithm $(\pi/2)/[1 + \delta\pi/2)]$.

δ	ϵ	$\pi/2 - \epsilon$	$(\pi/2)/[1 + \delta\pi/2)]$
0.00	0.00	1.57	1.57
0.01	0.05	1.52	1.55
0.05	0.18	1.39	1.46
0.10	0.30	1.27	1.36
0.20	0.47	1.10	1.20
0.40	0.68	0.89	0.96
0.80	0.92	0.65	0.70
1.00	1.00	0.57	0.61
1.50	1.13	0.44	0.47
2.00	1.21	0.36	0.38
4.00	1.36	0.21	0.22
8.00	1.46	0.11	0.12
10.00	1.48	0.09	0.09
20.00	1.52	0.05	0.05
100.00	1.56	0.01	0.01
∞	1.57	0.00	0.00

Equation (4.43) describes the daily increment to total dry weight of a crop in terms of the photosynthetic characteristics of upper unshaded leaves in the 'closed' crop canopy. The effects of environmental and genetic modifications of these characteristics have been examined in Chapter 2, and the mathematical consequences of these modifications on total dry matter production can be directly estimated. Careful examination of eqn (4.43) shows two components of total dry matter productivity. First there is the component due to the photosynthetic performances of the constituent leaves and second the component attributable to the proportion of incident light intercepted by the crop (the term $1 - \exp(-KL)$). We could write eqn (4.43) as

$$\Delta W/\Delta t = 0.273 Y \nabla_{\mathrm{MAX}} [1 - \exp(-KL)]/f_{\mathrm{C}}, \qquad (4.44)$$

where ∇_{MAX} is the daily photosynthetic integral at full light interception. We could also write eqn (4.44) as

$$\frac{\Delta C}{\Delta t} = f_{\mathrm{C}} \frac{\Delta W}{\Delta t} = 0.273 Y \nabla_{\mathrm{MAX}} [1 - \exp(-KL)]. \qquad (4.45)$$

4.5. CROP PHOTOSYNTHESIS AND PRODUCTIVITY

We are not usually concerned, either in agriculture or horticulture, with harvesting the entire dry matter yield of the plant or crop. Our interest is usually focussed on some particular component of the total dry matter. For instance, with grain crops such as wheat, rice or sorghum we are most interested in the component of seed yield, whilst with crops such as potato and sugar beet we are most interested in the component of 'root' yield. It is really only with forage crops that our interest is more nearly focussed on the total plant or crop dry matter yield. Simple predictions or analyses of crop *total* dry matter yield of the type described in Section 4.3 therefore are only partially successful in meeting our needs. We need to develop more complete physiological analyses of crop photosynthesis and productivity which distinguish between the different components of crop yield.

Let us consider a growing vegetative crop with two dry matter components, the root weight per unit ground area, R, and the shoot weight per unit ground area, T. If we relax the steady-state assumption that is implicit in the analysis derived in Section 4.2, we could write, using eqns (4.11) and (4.12), that

$$\frac{\Delta T}{\Delta t} = \frac{1}{f_C} \left[\frac{\sigma_N}{\sigma_N + \sigma_C(f_N/f_C)} \right] \sigma_C T. \qquad (4.46)$$

If we now define a partition coefficient η_T such that

$$\eta_T = \sigma_N / [\sigma_N + \sigma_C(f_N/f_C)] \qquad (4.47)$$

we can re-write eqn (4.46) as

$$\Delta T/\Delta t = \eta_T \sigma_C T/f_C \qquad (4.48)$$

and then, using eqn (4.4),

$$\Delta T/\Delta t = (\eta_T/f_C) \, \Delta C/\Delta t. \qquad (4.49)$$

We can add that the rate of root growth is given by

$$\Delta R/\Delta t = (\eta_R/f_C) \, \Delta C/\Delta t, \qquad (4.50)$$

where η_R is the root partition coefficient and

$$\eta_R + \eta_T = 1. \qquad (4.51)$$

If we add eqns (4.48) and (4.50) we have

$$\Delta T/\Delta t + \Delta R/\Delta t = \Delta W/\Delta t = [(\eta_R + \eta_T)/f_C] \, \Delta C/\Delta t,$$

and finally,

$$\frac{\Delta W}{\Delta t} = \frac{1}{f_C} \frac{\Delta C}{\Delta t}, \qquad (4.19)$$

where $\Delta W/\Delta t$ is the growth rate of the whole crop. For a crop with n distinct physiological components eqns (4.48) and (4.50) can be generalized as

$$\Delta W_n/\Delta t = (\eta_n/f_C)\,\Delta C/\Delta t, \qquad (4.52)$$

where $\Delta W_n/\Delta t$ is the growth rate of the nth component, and the partition coefficients η_n give rise to the expression

$$\sum_n \eta_n = 1. \qquad (4.53)$$

The number of physiological components, n, of the crop will change with the physiological state of the crop. Whilst it is in the vegetative state we will generally only be concerned with the change in dry weights of the leaves, W_L, the roots, W_R, and the stems, W_S. When the crop enters the reproductive growth state we will need to consider the partition or allocation, or newly assimilated dry matter to the reproductive parts of the crop, flowers and seeds etc., whose combined dry weight we can denote by W_F. It is therefore helpful to consider vegetative and reproductive growth separately.

A. Vegetative Growth

After an extended period of vegetative growth many crops exhibit a maximum value for their standing leaf weight per unit of ground area. This occurs when their rate of loss of leaf dry weight, through leaf senescence and abscission, becomes equal to their rate of production of new leaf material. It is helpful if we assume that physiologically determined tissue death, that is 'natural' tissue death as opposed to tissue death through pest or disease actions, is primarily confined to the leaf tissues. We have seen, in the previous chapters, that the decline in photosynthetic and respiratory activity of tomato leaves, which suggests the more general decline of their metabolic activity, can be attributed to the deteriorating light environment they experience *in situ* in a 'closed' crop canopy. If we now assume that the rate of loss of dry matter is dependent both on the absolute level of the crop's metabolic activity, the photosynthetic rate, and the extent of leaf shading within the crop canopy, we can write the rate of loss of leaf dry weight per unit of ground area, A, as

$$A = \gamma^*\,(\Delta C/\Delta t)\int_0^L (I_0/I)\,dL, \qquad (4.54)$$

where γ^* is a constant, I the downward light flux density at any particular horizon within the canopy and L the cumulative leaf area index of the canopy. We can now write

$$\frac{\Delta W_L}{\Delta t} = (\eta_L/f_C)\,\Delta C/\Delta t - A, \qquad (4.55)$$

and, if we use eqn (3.7) to describe I_0/I,

$$\frac{\Delta W_L}{\Delta t} = \{\,\eta_L/f_C - \gamma^*[\exp(KL)-1]/K\,\}\,\Delta C/\Delta t. \qquad (4.56)$$

If, as a first approximation, we use the first term of the expansion of $\exp(KL) - 1$, eqn (4.56) simplifies to

$$\frac{\Delta W_L}{\Delta t} = (\eta_L/f_C)\,(1 - \gamma W_L)\,\Delta C/\Delta t, \qquad (4.57)$$

where

$$\gamma = \gamma^* f_C s_A/\eta_L \qquad (4.58)$$

and s_A is the specific leaf area $(=L/W_L)$.

During vegetative growth we could now write the rates of leaf, stem and root growth as

$$\Delta W_L/\Delta t = (\eta_L/f_C)\,(1 - \gamma W_L)\,\Delta C/\Delta t \qquad (4.59)$$

$$\Delta W_S/\Delta t = (\eta_S/f_C)\,\Delta C/\Delta t \qquad (4.60)$$

$$\Delta W_R/\Delta t = (\eta_R/f_C)\,\Delta C/\Delta t, \qquad (4.61)$$

where

$$\eta_L + \eta_S + \eta_R = 1. \qquad (4.62)$$

B. Reproductive Growth

For simplicity we will consider only a crop which is, in an agricultural sense, determinate. That is a crop which ceases to produce new leaf and stem dry weight at flowering. Some plants which are botanically determinate, i.e. they produce their flowers on the terminal meristem, continue to produce new leaves on axilliary meristems, and are agriculturally indeterminate, for example tomato plants. In contrast, some plants which are botanically indeterminate, i.e. they produce their flowers on axilliary meristems, experience a major check to the growth of the terminal meristem and are agriculturally determinate, for example the fibre crop kenaf *(Hibiscus cannabinus)*. The relevant assumptions can be made to deal with any particular crop, and the simplest situation is examined here.

At the start of flowering leaf production ceases, although canopy photosynthesis and leaf death continue, and eqn (4.59) can be written as

$$\Delta W_L/\Delta t = -\,(\gamma W_L \eta_L/f_C)\,\Delta C/\Delta t. \qquad (4.63)$$

Stem growth ceases, so that

$$\Delta W_S/\Delta t = 0, \qquad (4.64)$$

and if we assume that a proportion β of the dry matter that was going to produce new leaves and stems now goes to the reproductive parts

$$\Delta W_F = [\beta(\eta_L + \eta_S)/f_C]\, \Delta W/\Delta t. \qquad (4.65)$$

The rate of root growth now becomes

$$\Delta W_R/\Delta t = \{[\eta_R + (1 - \beta)(\eta_L + \eta_S)]/f_C\}\, \Delta C/\Delta t. \qquad (4.66)$$

Crudely, the constant β represents the investment of the crop in reproductive development and the proportion $1 - \beta$ its investment in perennial growth through storage of assimilate in the roots.

C. Crop Growth

We have now developed a simple model for crop growth. If we denote the flowering time as t_f, and use eqn (4.45) to substitute for $\Delta C/\Delta t$, we can write that for $t < t_f$,

$$\frac{\Delta W_L}{\Delta t} = \frac{0.273}{f_C}\, \eta_L Y \nabla_{MAX}(1 - \gamma W_L)[1 - \exp(-Ks_A W_L)] \qquad (4.67)$$

$$\frac{\Delta W_S}{\Delta t} = \frac{0.273}{f_C}\, \eta_S \nabla_{MAX}[1 - \exp(-Ks_A W_L)] \qquad (4.68)$$

$$\frac{\Delta W_F}{\Delta t} = 0 \qquad (4.69)$$

$$\frac{\Delta W_R}{\Delta t} = \frac{0.273}{f_C}\, \eta_R \nabla_{MAX}[1 - \exp(-Ks_A W_L)] \qquad (4.70)$$

and for $t > t_f$,

$$\frac{\Delta W_L}{\Delta t} = -\frac{0.273}{f_C}\, \eta_L Y \nabla_{MAX} \gamma W_L[1 - \exp(-Ks_A W_L)] \qquad (4.71)$$

$$\frac{\Delta W_S}{\Delta t} = 0 \qquad (4.72)$$

$$\frac{\Delta W_F}{\Delta t} = \frac{0.273}{f_C}\, \beta(\eta_L + \eta_S)Y \nabla_{MAX}[1 - \exp(-Ks_A W_L)] \qquad (4.73)$$

$$\frac{\Delta W_R}{\Delta t} = \frac{0.273}{f_C}\, [(1 - \beta)(\eta_L + \eta_S) + \eta_R] Y \nabla_{MAX}[1 - \exp(-Ks_A W_L)]. \qquad (4.74)$$

We can now add the relationship

$$X(t_i) = X(t_{i-1}) + \Delta X(t_{i-1}) \qquad (4.75)$$

where X represents any of the dependent variables W_L, W_S, W_F or W_R.

Equation (4.67) gives rise to the numerical estimate of the maximum standing leaf weight per unit of ground area $(W_L)_{MAX}$ as

$$(W_L)_{MAX} = 1/\gamma = \eta_L/\gamma^* f_C s_A . \tag{4.76}$$

Equations (4.67)–(4.74) can only be solved by numerical methods, using eqn (4.75), but with the appropriate approximation we can obtain analytical solutions for W_L, W_S, W_F and W_R. If we make the approximation

$$1 - \exp(-K s_A W_L) \cong K s_A W_L \tag{4.77}$$

and write

$$\mu_L = \frac{0.273}{f_C} \eta_L Y \nabla_{MAX} K s_A \tag{4.78}$$

$$\mu_S = \mu_L \eta_S/\eta_L \tag{4.79}$$

$$\mu_R = \mu_L \eta_R/\eta_L \tag{4.80}$$

we can then show that

$$W_L = W_{L_0} \exp(\mu_L t)/\{1 - \gamma L_0 [1 - \exp(\mu_L t)]\}. \tag{4.81}$$

$$W_S = W_{S_0} + \mu_S t/\gamma + (\mu_S/\mu_L \gamma) \ln(W_{L_0}/W_L) \tag{4.82}$$

$$W_R = W_{R_0} + \mu_R t/\gamma + (\mu_R/\mu_L \gamma) \ln(W_{L_0}/W_L) \tag{4.83}$$

$$W_F = 0, \tag{4.84}$$

for $t < t_f$, and for $t > t_f$

$$W_L = W_{L*} /[1 + \gamma W_{L*} \mu_L (t - t_f)] \tag{4.85}$$

$$W_S = W_{S*} \tag{4.86}$$

$$W_R = W_{R*} \{ [\mu_R + (1 - \beta)(\mu_S + \mu_L)]/\mu_L \gamma \} \ln(W_{L*}/W_L) \tag{4.87}$$

$$W_R = [\beta(\mu_S + \mu_L)/\mu_L \gamma] \ln(W_{L*}/W_L), \tag{4.88}$$

where W_{L0}, W_{S0} and W_{R0} are leaf, stem and root dry weights at $t = 0$ and W_{L*}, W_{S*} and W_{R*} their dry weights at $t = t_f$.

Now eqn (4.85) predicts that after flowering the leaf mass of the crop approaches zero asymptotically with time, and the cumulative weight of the reproductive parts increases infinitely (eqn (4.88)). This latter behaviour is non-physiological and non-agricultural. If we assume that all effective carbon assimilation ceases when the leaf mass per unit ground area falls below some critical level W_{LC}, the maximum yield of reproductive parts $(W_F)_{MAX}$, can be written as

$$(W_F)_{MAX} = [\beta(\mu_S + \mu_L)/\mu_L \gamma] \ln(W_{L*}/W_{LC}). \tag{4.89}$$

Despite the deficiency of eqn (4.88) the analytical solutions enable us to obtain some direct feeling of the growth pattern predicted by the analysis.

4.6. AN ANALYSIS OF CROP GROWTH DATA

With initial estimates of the parameters of eqns (4.67)–(4.74) crop growth can be simulated over any time period. Values of the predicted dry weights of any, or all, of the crop components can be abstracted from the simulated data set and compared with real experimental data. Using suitable computer techniques the parameters can be adjusted so as to minimize the residual sum of the squares of the differences between the predicted and real data values. Simply, we can 'fit' the model to crop growth data in exactly the same way as was discussed in Section 1.2. The only difference between this and other models described before is that it needs to be solved by the techniques of numerical integration.

Many of the parameters, eqns (4.67)–(4.74) contain eleven parameters (i.e. η_L, η_S, η_R, Y, ∇_{MAX}, f_C, K, s_A, γ, β, t_f), appear in association, one with the other, and for analysis of field data the model can be simplified to

$$\frac{\Delta W_L}{\Delta t} = a_L(1 - \gamma W_L)[1 - \exp(-bW_L)], \qquad t \leqslant t_f \qquad (4.90)$$

$$\frac{\Delta W_S}{\Delta t} = a_S[1 - \exp(-bW_L)], \qquad t \leqslant t_f \qquad (4.91)$$

$$\frac{\Delta W_F}{\Delta t} = 0, \qquad t < t_f \qquad (4.92)$$

$$\frac{\Delta W_R}{\Delta t} = a_R[1 - \exp(-bW_L)], \qquad t \leqslant t_f \qquad (4.93)$$

and

$$\frac{\Delta W_L}{\Delta t} = -\gamma a_L W_L[1 - \exp(-bW_L)], \qquad t > t_f \qquad (4.94)$$

$$\frac{\Delta W_S}{\Delta t} = 0, \qquad t > t_f \qquad (4.95)$$

$$\frac{\Delta W_F}{\Delta t} = \beta(a_L + a_S)[1 - \exp(-bW_L)] \qquad (4.96)$$

$$\frac{\Delta W_R}{\Delta t} = [a_R + (1 - \beta)(a_L + a_S)][1 - \exp(-bW_L)], \qquad (4.97)$$

where

$$a_L = 0.273\eta_L Y\nabla_{MAX}/f_C \qquad (4.98)$$

$$a_S = a_L\eta_S/\eta_L \qquad (4.99)$$

$$a_R = a_L\eta_R/\eta_L \qquad (4.100)$$

$$b = Ks_A. \tag{4.101}$$

Equations (4.90)–(4.97) now contain seven parameters (i.e. a_L, a_S, a_R, b, γ, β and t_f).

The analysis has been applied to harvest data for a field grown crop of the forage legume *Stylosanthes humilis*. For this crop there were regular harvest values for the standing weights of leaf, stem and reproductive tissues and independent estimates of the median crop flowering times, t_f. Equations (4.90)–(4.92) and (4.94)–(4.96) are 'fitted' to these harvest data after their numerical integration. To facilitate the fitting procedures the sum of squares of the differences between the square roots of the predicted and observed data values was minimized.

TABLE 4.4. (a) Numerical estimates of the five coefficients (a_L, a_S, b, γ and β) obtained by 'fitting' eqns (4.90)–(4.92) and eqns (4.94)–(4.96) to field data for the growth of the legume *Stylosanthes humilis*. (b) Correlation matrix for the five coefficients after 'fitting' the model to the experimental data.

<table>
<tr><td colspan="6" align="center">(a)</td></tr>
<tr><td>a_L</td><td colspan="5">7.0 (±1.0) g d^{-1}</td></tr>
<tr><td>a_S</td><td colspan="5">7.7 (±0.9) g d^{-1}</td></tr>
<tr><td>b</td><td colspan="5">0.017 (±0.033) m^2 g^{-1}</td></tr>
<tr><td>γ</td><td colspan="5">0.0046 (±0.0013) m^2 g^{-1}</td></tr>
<tr><td>β</td><td colspan="5">0.31 (±0.17)</td></tr>
</table>

	a_L	a_S	b	γ	β
			(b)		
a_L	1.00				
a_S	0.98	1.00			
b	−0.99	−0.99	1.00		
γ	−0.41	−0.47	0.45	1.00	
β	−0.92	−0.92	0.92	0.51	1.00

The estimated values of the five coefficients, a_L, a_S, b, γ and β, together with their approximate 95% confidence intervals, for one set of the field data are shown in Table 4.4. The correlation matrix for the five parameters is also shown in Table 4.4. The fit of the model to the data is illustrated in Fig. 4.4. Whilst the model provides a good visual description of the real changes in dry weight of the crop components with time the high correlation between the coefficients a_L and b and a_S and b is disturbing. The numerical value of b was also so poorly determined that it needs to be treated with some scepticism. The physiological quantities K (canopy extinction coefficient) and s_A (specific leaf area) are amenable to simple and direct measurement in the field. With an independent estimate of b (Ks_A) of 0.0169 m^2 g^{-1} the remaining four parameters were

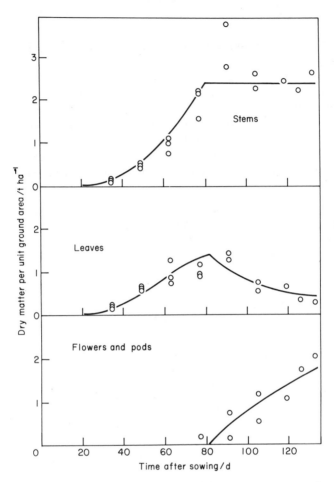

FIG. 4.4. The 'fit' of eqns (4.90)–(4.92) and eqns (4.94)–(4.96) to field data for the legume *Stylosanthes humilis*.

re-estimated. The new estimates, together with the approximate 95% confidence intervals, and the parameter correlation matrix is shown in Table 4.5. It is worthwhile examining the original objectives of the field experiment before looking at the results of the complete analysis of the experimental data.

S. humilis is a self-seeding annual pasture legume grown commercially over much of northern Australia to supplement or replace the low protein native pasture species available for cattle feed. In those areas where it is grown there are wide variations, both spatial and temporal, in the date of the germinating rains and the length of the following growing season. There is also considerable variation in the flowering behaviour of different ecotypes, which would affect

TABLE 4.5 (a) Numerical estimates of the four coefficients (a_L, a_S, γ and β) obtained by 'fitting' eqns (4.90)–(4.92) and eqns (4.94)–(4.96) to field data for the growth of *Stylosanthes humilis*. The parameter b was constrained to 0.017 m^2 g^{-1}. (b) Correlation matrix for the four coefficients after 'fitting' the model to the experimental data.

(a)

a_L	7.1 (±1.0) g d^{-1}
a_S	7.5 (±1.0) g d^{-1}
γ	0.0049 (±0.0012) $m^2 g^{-1}$
β	0.31 (±0.07)

(b)

	a_L	a_S	γ	β
a_L	1.00			
a_S	−0.67	1.00		
γ	−0.38	−0.09	1.00	
β	−0.02	0.29	−0.09	1.00

the relative lengths of the periods of vegetative and reproductive growth. The interaction between length of the growing season and the relative lengths of vegetative and reproductive growth periods will affect both absolute and relative yields of forage and seed. The interaction is clearly of agronomic interest.

Accordingly, three ecotypes of *S. humilis*, with contrasting flowering behaviour, were grown under irrigation during the wet summer season at the Katherine Research Station, Northern Territory, Australia. Growing seasons of different lengths were obtained by sowing crops of each ecotype at 4 week intervals. Samples were harvested from each of the crops at 2 week intervals during both vegetative and reproductive growth.

Equations (4.90)–(4.92) and (4.94)–(4.96) were fitted to the harvest data for each crop. A value of 0.017 g m^{-2} was assumed for the parameter b, and t_f was obtained from independent estimates of the median flowering time of each crop. Estimates of the four parameters a_L, a_S, γ and β for the three sowings of each of the three ecotypes are shown in Table 4.6. The fitted growth curves and harvest data for the three sowings of the mid-flowering ecotype Katherine are shown in Fig. 4.5.

The only systematic differences in the estimates of the parameters are the lower values of the parameter a_L obtained from the analysis of the December sowing of all three ecotypes. There being no other systematic differences in the numerical values of the 'fitted' parameters we can have some confidence in concluding that, at the level of resolution available for these data, the three

TABLE 4.6. Numerical estimates of the four coefficients (a_L, a_S, γ and β), and independent estimates of the time between sowing and flowering, t_f, obtained by fitting eqns (4.90)–(4.92) and (4.94)–(4.96) to field data for the growth of three ecotypes of the legume *Stylosanthes humilis* each sown at three times of the year. (G, ecotype Greenvale; Kt, ecotype Katherine; KL, ecotype Kalumburu; N, November sowing; D, December sowing; J, January sowing.)[a]

Ecotype	Sowing	a_L/g d^{-1}	a_S/g d^{-1}	γ/10^{-3} m^2 g^{-1}	β	t_f/d
G	N	10.9 (1.7)	6.3 (0.5)	4.9 (0.7)	0.33 (0.08)	115
G	D	7.1 (1.0)	7.5 (1.0)	4.9 (1.2)	0.31 (0.07)	80
G	J	8.8 (1.3)	7.1 (1.1)	6.0 (1.4)	0.44 (0.08)	65
Kt	N	10.4 (1.4)	7.4 (0.5)	5.3 (0.6)	0.26 (0.06)	116
Kt	D	7.0 (0.8)	8.9 (1.0)	4.4 (0.9)	0.24 (0.05)	84
Kt	J	9.0 (1.1)	9.7 (1.1)	3.4 (0.8)	0.31 (0.05)	67
KL	N	10.2 (1.7)	6.6 (0.6)	6.1 (0.8)	0.30 (0.12)	133
KL	D	7.4 (1.2)	7.6 (0.9)	4.9 (1.1)	0.21 (0.08)	105
KL	J	10.6 (1.3)	7.1 (0.6)	5.3 (0.8)	0.14 (0.11)	98

[a] In all cases the parameter b was constrained to 0.017 m^2 g^{-1}.

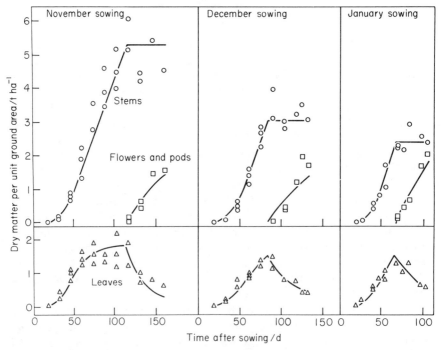

FIG. 4.5. Experimental harvest data and 'fitted' growth curves for the *Stylosanthes humilis* ecotype Katherine sown at three times of the year (November, December and January).

ecotypes are physiologically similar in all respects other than in their flowering behaviour. Since the crops were irrigated, supplied with adequate fertilizer and the daily incident total radiation integral varied between 20 and 25 MJ m^{-2} it is not unreasonable to suppose that the parameters ∇_{MAX}, a_L and a_S were fairly constant during the entire growth period of each crop.

We made the assumption that after flowering only a proportion β of the daily dry matter increment previously partitioned to the shoot (leaves and stems) was partitioned to the reproductive parts. The remainder, $1 - \beta$, was presumed to be stored in the roots as a potential source of substrate for perennial growth. If this assumption is correct β is an index of the relative investment of the crop in reproductive, compared with perennial, growth. For *S. humilis* β was estimated to be about 0.3, that is 30% of the current assimilate normally partitioned to the shoot during vegetative growth is, after flowering, partitioned to the reproductive parts. There were no data on root dry weights per unit ground area, so that it is not possible to test whether, after flowering, the other 70% was partitioned to the roots. However, it has been observed that when grown under irrigation the crop does exhibit some perennial character-

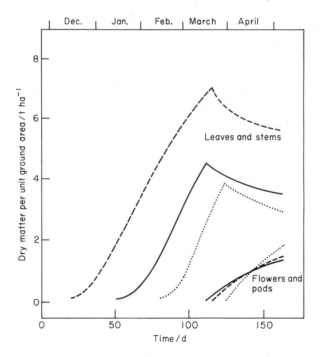

FIG. 4.6. 'Fitted' growth curves for the forage component (leaves and stems) and reproductive component (pods, seeds, etc.) of the *Stylosanthes humilis* ecotype Katherine sown at three times of the year.

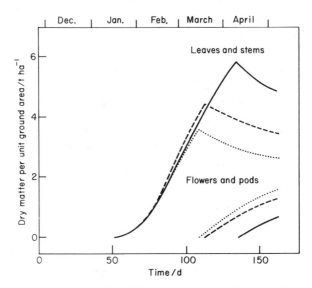

FIG. 4.7. 'Fitted' growth curves for the forage components (leaves and stems) and reproductive component of the three ecotypes of *Stylosanthes humilis* sown at the same time
(solid line, Kalumburu; broken line, Katherine; dotted line, Greenvale).

istics. The parameter β is clearly of some ecological and physiological interest.

The analysis suggests that the main physiological differences between the
three ecotypes (the early flowering 'Greenvale', mid-flowering 'Katherine' and
late-flowering 'Kalumburu') were attributable to their differences in flowering
behaviour. This is illustrated by Figs 4.6 and 4.7 where the predicted harvest data
for forage yield (leaf and stem dry weights) and the yield of reproductive parts
are plotted against the time after the sowing of the first crop. In Fig. 4.6 the
figures show that despite the different sowing dates for the three crops of the
ecotype 'Katherine' they all flowered at approximately the same time. However, the longer the period of vegetative growth the greater the forage yield,
although the yield of reproductive parts was not greatly affected. In Fig. 4.7 are
shown the predicted growth curves of the three ecotypes sown at the same
time. The late-flowering ecotype 'Kalumburu' had a greater forage yield, but
lower yield of reproductive parts, whilst the exact opposite occurred with the
early-flowering ecotype 'Greenvale''. The growth curves also show the similarity
of the three ecotypes in their vegetative growth behaviour.

Often less comprehensive data sets are available than those described above.
However, the analysis can still be usefully applied to them. For example, total
shoot dry weight and crop leaf area indices have been recorded for crops of the
fibre plant 'kenaf' (*H. cannabinus*) during successive sowings at the Kimberley
Research Station in Western Australia. If we assume that $\beta = 1$, addition of both

eqns (4.90) and (4.91) and eqns (4.94), (4.95) and (4.96) yield

$$\frac{\Delta T}{\Delta t} = (a_L + a_S)[1 - \exp(-KL)] - a_L \gamma W_L[1 - \exp(-KL)], \quad (4.102)$$

for both $t \leqslant t_f$ and $t > t_f$, where T is the shoot dry weight per unit ground area and L the leaf area index of the crop. Equation (4.102) can be written in the simpler form

$$\frac{\Delta T}{\Delta t} = a_T[1 - \exp(-KL)] - qT, \quad (4.103)$$

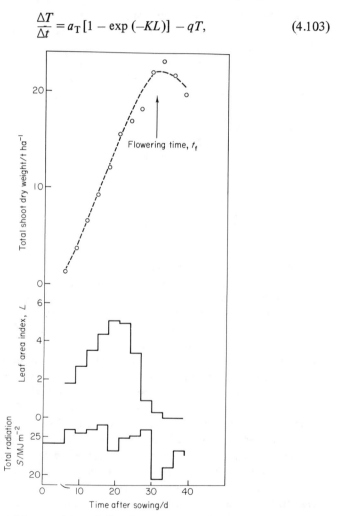

FIG. 4.8. Experimental harvest data and 'fitted' growth curve for the shoot dry weights of *Hibiscus cannabinus*. Also shown are the 'average' values of the leaf area index and daily radiation integrals between harvests.

where q is a constant describing loss of shoot dry weight. Using average values of the measured leaf area index for the periods between harvests eqn (4.103) can be 'fitted' to the shoot dry weight data. For the 'kenaf' crop a value of $K = 0.5$ was assumed throughout the growth period and the parameters a_T and q are adjusted to provide the best 'fit' to the experimental data. The average values of leaf area index together with measured shoot dry weights and 'fitted' growth curve for one sowing of the 'kenaf' crop are shown in Fig. 4.8. Also shown are the average daily radiation integrals between harvests. The optimized values of a_T and γ were 21.0 g m^{-2} d^{-1} and 4.1 x 10^{-3} d^{-1}. The value of a_T for the kenaf crop can be compared with the average value of $(a_L + a_S)$ of about 16.6 g m^{-2} d^{-1} for the crops of *S. humilis* grown under similar radiation levels. The difference could be attributed to differences in a_R, or, since a_L and a_S are potentially highly correlated with K (see Table 4.4), to the assumption that K has the particular value 0.5.

Equation (4.103) serves to illustrate one very important aspect of crop growth. The growth rate is both a function of the potential photosynthetic activity of the canopy *and* the proportion of the incident radiation intercepted by the crop.

4.7. CONCLUSIONS

Traditionally, studies of crop growth have been based on the analysis of sequential harvest data, using polynomial or exponential equations to describe changes in crop weight as a direct function of the time variable. Whilst this type of analysis has played an important role in the development of our quantitative understanding of crop growth, it has a number of notable deficiences. Since the mathematical equations used in the primary analysis of the crop data are empirical their coefficients have little, if any, direct physiological meaning. The crop attributes derived from the coefficients of these equations, attributes such as relative growth rates and net assimilation rates, are dependent on the physiological state of the crop. Comparison of the behaviour of crops subjected to different growth environments is therefore on the basis of somewhat arbitrary coefficients or state-dependent attributes, neither basis being particularly satisfactory. The primary analysis cannot effectively accommodate changes in the crop's environment during growth. Such changes are usually dealt with by retrospective analysis of the correlation between derived crop attributes and the appropriate environmental variables. These deficiencies are worsened if the crops change from vegetative to reproductive growth during the experimental period.

In this chapter I have developed a mechanistic approach to the mathematical analysis of crop growth data which, to a large extent, overcomes these deficiences. It has been shown in Section 4.6 that the analysis can, using computer techniques, be readily applied to experimental data obtained for field crops. The coefficients derived in the primary analysis are physiologically defined, and the analysis can accommodate daily changes in the crop's environment during growth. Although the mechanistic basis of the analysis is not rigorous, it is based on a set of equations that would provide an adequate framework for a simulation model of crop growth.

The 'goodness of fit' of this type of model to real field data must lend some credence to its predictions of yield when used in a simulation mode. For example, the agriculturist or horticulturist who either breeds new plant varieties or seeks to introduce new plant species is interested in identifying those plant characteristics which, in any particular growth environment, determine the potential harvestable yield of the economic component of the crop. Now, growth is an emergent property of the crop. That is, growth results from the interactions of the primary nutrient assimilating process (cf. Section 4.2) and their interactions with the physical structure of the crop and its aerial and root environments. It is quite possible that a particular characteristic which profoundly affects the rate/or extent of crop growth in one environment may have little effect in another. The simplest approach to adopt is to grow all new plant material from seedling to harvestable stage in different growth environments, and to compare yields directly. This approach, however, is demanding on time and other resources. Mathematics, and more specifically mathematical models, provide a useful tool to study the many potential interactions of plant characteristics. Now models, as we have seen, are simplified descriptions of real systems; they represent solutions of the formal statements of the assumptions that have been made about the real systems. If a model has been shown to provide a 'good description' of real field data then it is not unreasonable to expect that it will provide a sensible indication of real effects of differences in plant characters, simulated by suitable changes in the model's parameters, on crop growth.

The thesis can be illustrated by using eqns (4.67)–(4.76) as the basis for a simple simulation model of wheat growth under 'average' U.K. weather conditions. For a determinate crop, such as wheat, the production of new leaves ceases when the flower is initiated on the terminal meristem. Whilst the expansion of existing leaves, and stem elongation, continue for some time after flower initiation it is not unreasonable to suppose that at, or about, anthesis all vegetative growth ceases, and thereafter all newly acquired dry matter is allocated to the reproductive tissues. If we assume that the only significant loss of shoot material through physiological causes is the loss of leaf material through senescence and abscission, the daily increment in leaf dry weight, ΔW_L, can be written as

$$\Delta W_L = \frac{0.273}{f_c} \eta_L \nabla_{MAX} \left[1 - \exp(-K s_A W_L) \right] - \nu, \qquad (4.104)$$

where ν is the daily loss of leaf dry matter through senescence etc., and the other parameters are as previously defined. The daily increments in stem, ΔW_S, and root, ΔW_R, dry weights are as given in eqns (4.68) and (4.70), respectively. We need to define ν. If we assume that the daily loss of leaf dry matter is directly proportional to the standing shoot mass we can write

$$\nu = \gamma^*(W_L + W_S), \qquad t < t_a, \qquad (4.105)$$

where γ^* is a constant and t_a the anthesis date (expressed as days after crop emergence). After anthesis all newly acquired dry matter is partitioned to the reproductive parts, and so,

$$\Delta W_L = -\nu, \qquad t > t_a,$$
$$\Delta W_S = \Delta W_R = 0, \qquad t > t_a, \qquad (4.106)$$

and

$$\Delta W_F = \frac{0.273}{f_c} \nabla_{MAX} [1 - \exp(-Ks_A W_L)], \qquad t > t_a. \qquad (4.107)$$

The daily decrement in leaf dry weight then becomes,

$$\nu = \gamma^*(W_L + W_S + W_F), \qquad t > t_a. \qquad (4.108)$$

We can also write that

$$\nabla_F = \nabla_{MAX} [1 - \exp(-Ks_A W_L)]/Y, \qquad (4.109)$$

where ∇_F is given by eqn (4.42), whence ∇_{MAX} becomes

$$\nabla_{MAX} = Y\alpha S\tau_0 CZ/[K\alpha S + (1 - m)\tau_0 CZ]. \qquad (4.110)$$

The daily canopy photosynthetic integral, ∇_F, is written in eqn (4.109) as a function of the leaf area index of the crop (i.e. $s_A W_L = L$). We can obtain the daily increment in leaf area index as $s_A \Delta W_L$. However, the specific leaf area, s_A, is a function of the crop's growth environment. For both tomato and chrysanthemum plants we know that as a first approximation

$$s_A = H/(a_0 H + a_1 SH + a_2), \qquad (4.111)$$

where H is the average 'bulk air' temperature. Equation (4.111) could be used to calculate the daily increment in leaf area index for a particular increment in standing leaf dry weight. Note that we can also replace τ_0 by τ_0^*/s_A, thereby allowing for seasonal changes in leaf photosynthetic activity due to changes in leaf thickness. As leaves die the older ones will presumably die first. Since s_A may change with time of year the leaf area lost for a given loss in standing leaf weight may also change with time of year. It will therefore be important to keep a record of daily increments in both leaf mass and area to ensure that the oldest leaves are lost first, and the leaf area index changes are kept correctly.

TABLE 4.7. Monthly averages of the daily light integral, mean daylength and mean air temperature for the town of Aberystwyth, Dyfed, U.K.

Month	Daily light integral, $S/MJ\ m^{-2}$	Mean daylength, Z/s	Mean air temperature, $H/^{\circ}C$
January	1.0	3.24×10^4	4.3
February	2.1	3.82×10^4	4.6
March	4.0	4.39×10^4	6.1
April	6.2	5.08×10^4	8.0
May	7.7	5.72×10^4	11.0
June	8.7	6.08×10^4	13.7
July	7.8	5.94×10^4	15.7
August	6.7	5.33×10^4	15.7
September	4.8	4.75×10^4	13.5
October	2.7	4.10×10^4	10.9
November	1.4	3.42×10^4	7.2
December	0.8	3.24×10^4	4.3

It now only remains to add the relationship given by eqn (4.75), that is

$$X(t_i) = X(t_{i-1}) + \Delta X(t_{i-1}), \qquad (4.75)$$

to enable the model to be numerically integrated on a day-to-day basis. This is very simply done by computer.

Monthly averages of the daylength, daily light integral and mean air temperature for the town of Aberystwyth, Dyfed, U.K. are shown in Table 4.7. Using these data it is possible to use the model to predict grain yield for a wheat crop

TABLE 4.8. Parameter values used for the simulations of wheat productivity reported in Section 4.7.

α	$11.5 \times 10^{-6}\ g\ J^{-1}$
τ_0^*	$3.3 \times 10^{-5}\ m^2\ g^{-1}\ s^{-1}$
Y	0.6
K	0.6
m	0.125
f_C	0.36
η_L	0.4
η_S	0.4
a_0	$-6.9\ g\ m^{-2}$
a_1	$9.2 \times 10^{-6}\ g\ J^{-1}$
a_2	$42.9\ g\ m^{-2}\ (^{\circ}C)$
α	$0.005\ g\ g^{-1}\ d^{-1}$

Initial values of W_L and W_S at the nominal emergence date

W_{L_0}	$0.1\ t\ ha^{-1}$
W_{S_0}	$0.1\ t\ ha^{-1}$

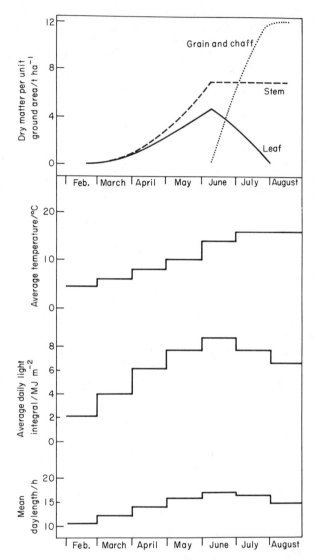

FIG. 4.9. Simulated growth of a wheat crop under U.K. conditions. The nominal emergence and anthesis dates were assumed to be 19 February and 9 June, respectively. The crop reached maturity on 2 August. Also shown are the presumed 'average' U.K. meteorological data.

planted, and reaching anthesis, at different times of the year. It is assumed that 87.5% of the cumulated reproductive yield, following anthesis, is grain. Using the parameter values shown in Table 4.8, together with the initial dry weights of

4. Photosynthesis and Productivity

leaf, stem and root tissues shown in the table, and the meteorological data given in Table 4.7, growth can be simulated over any desired time period. For example, growth of a crop, attaining the initial standing leaf, stem and root weights on 19 February, is shown in Fig. 4.9. It was assumed that this crop reached anthesis on 9 June. Following the simple way in which the model is set up all growth ceases when the standing leaf weight, or perhaps one should say the photosynthetically effective leaf weight, falls to zero. It is convenient to denote the time of initial standing leaf, stem and root weights of 0.1, 0.1 and 0.05 t ha^{-1} by t_e, and call t_e the emergence date. It is also convenient to denote the time at which reproductive growth ceases, that is when the standing leaf mass becomes zero, by t_m, and we can call t_m the maturity date.

We can use the meteorological data given in Table 4.7 and the parameter values in Table 4.8 to explore the effects on grain yield of selection of genotypes with different rates of light-saturated leaf photosynthesis, F_{MAX}. For example, if differences in F_{MAX} arise solely as a result of differences in leaf thickness (see Section 2.10) the simulations suggest an optimum value of s_A and F_{MAX} for maximum grain yield. This is because although thicker leaves (decreasing s_A) lead to higher values of F_{MAX}, the rate of leaf expansion, and

TABLE 4.9. The simulated effects on productivity of selection of wheat genotypes on the basis of their rates of light-saturated photosynthesis. The nominal emergence date of all crops was 17 February and nominal anthesis date was 28 May. L_A denotes the leaf area index at anthesis, W_S and W_G denote stem and grain dry weights at maturity. All crops reached maturity on 22 or 23 July. (a) Effects due to differences in the leaf carboxylation constant, τ_0^*. (b) Effects due to differences in specific leaf area, s_A.

(a)

$F_{MAX}/10^{-3}$ g m^{-2} s^{-1}	$\tau_0^*/10^{-5}$ m^3 g^{-1} s^{-1}	L_A	W_S/t ha^{-1}	W_G/t ha^{-1}
0.76	1.67	3.0	2.9	5.2
1.15	2.50	4.8	4.6	8.1
1.53	3.33	6.0	5.8	10.2
1.91	4.17	7.0	6.6	11.6
2.29	5.00	7.6	7.3	12.8

(b)

$F_{MAX}/10^{-3}$ g m^{-2} s^{-1}	$s_A/10^{-2}$ m^2 g^{-1}	L_A	W_S/t ha^{-1}	W_G/t ha^{-1}
0.76	2.62	9.5	4.7	7.9
1.01	1.98	8.2	5.3	9.0
1.53	1.31	6.0	5.8	10.2
2.04	0.98	4.4	5.6	10.2
2.29	0.66	2.2	4.1	7.9

the amount of intercepted light per unit ground area at anthesis, decrease as leaf thickness increases. However, if F_{MAX} is increased by selection of genotypes with high carboxylation constants, τ_0, but the same leaf thicknesses, there is a good correlation between F_{MAX} and grain yield. The results of these simulations are shown in Table 4.9. Selection on the basis of F_{MAX} alone may not, therefore, be a good breeding criterion. It would be more sensible to select genotypes for high carboxylation constants, τ_0, and thin leaves, that is low s_A.

Again, canopy architecture will affect the crop's photosynthetic performance. The effects of selecting genotypes on the basis of their canopy architecture are demonstrated in Table 4.10. In general, the genotypes with more erect leaves will form canopies with lower extinction coefficients. The results of the simulations shown in Table 4.10 suggest that there is an optimum canopy extinction coefficient, of about 0.5, and thereby an optimum canopy architecture.

It was pointed out in Section 1.3 that photosynthesis was only one of a number of processes by which plants acquired materials from their external environment. Other assimilating processes, and the timing of developmental processes, will affect crop yield. This is demonstrated by the results shown in Table 4.11. In the first part of the table the date of 'emergence' (t_e) is varied whilst the nominal anthesis date is maintained constant at 28 May. Winter sowing, that corresponds to the 'emergence' dates October–December, does not materially increase yield compared with the spring sowing, corresponding to the 'emergence' date around February. However, if spring sowing is delayed and the crop does not 'emerge' until March, there is a marked decrease in grain yield. The simulations suggest, therefore, that winter sowings do not intrinsically lead to higher yields, but ensure a more stable crop. In the second part of the table the effects of different anthesis dates for a spring sown crop are shown. There is apparently an optimum period between emergence and anthesis. Whereas the 'emergence' date does not affect the time of crop maturity, the longer the

TABLE 4.10. The simulated effects on productivity of selection of wheat genotypes on the basis of their canopy architecture. The nominal emergence date of all crops was 17 February and nominal anthesis date was 28 May. K denotes canopy extinction coefficient, L_A the leaf area index at anthesis, and W_S and W_G the stem and grain yields at maturity. All crops reached maturity on 23 July.

K	L_A	$W_S/t\ ha^{-1}$	$W_G/t\ ha^{-1}$
0.4	5.6	5.3	9.9
0.5	6.0	5.7	10.3
0.6	6.0	5.8	10.2
0.7	5.9	5.7	9.9
0.8	5.6	5.5	9.5

TABLE 4.11. The simulated effects of different emergence dates (t_e) and anthesis dates (t_a) on the productivity of wheat grown under U.K. conditions. (a) Effects due to differences in the emergence date with a nominal anthesis date of 28 May. In all simulations the maturity date (t_m) was between 18 and 22 July. (b) Effects due to differences in the anthesis date with a nominal emergence date of 19 February. (L_A denotes the leaf area index at anthesis, W_S and W_G the stem and grain yields and t_m the maturity date.)

(a)

t_e	L_A	W_S/t ha^{-1}	W_G/t ha^{-1}
20 October	6.9	8.2	10.0
19 November	7.0	7.8	10.2
19 December	6.9	7.4	10.4
18 January	6.8	6.8	10.5
17 February	6.0	5.6	10.1
19 March	4.0	3.6	8.5

(b)

t_a	L_A	t_m	W_S/t ha^{-1}	W_G/t ha^{-1}
20 May	5.4	12 July	4.8	9.5
30 May	6.0	23 July	5.8	10.2
9 June	6.6	3 August	6.9	10.4
19 June	7.1	15 August	8.1	10.4
29 June	7.6	26 August	9.3	10.2
7 July	8.1	5 September	10.3	9.8

period between 'emergence' of the spring crop and anthesis the longer the time between anthesis and maturity. Whilst the predicted period of grain filling is about 50 d for the crops reaching anthesis in late May–early June, it extends to about 60 d for crops reaching anthesis in late June–early July.

These simulations predict grain yields of about 10 t ha^{-1}, which is not unreasonable for the theoretical yields of wheat crops grown in the U.K. It should be noted that the parameter values shown in Table 4.8 are based on measurements of chrysanthemum and tomato crops. The simulations provide some support for the thesis that the principal differences between plant species lie in their morphological and developmental patterns, not in their utilization of the sun's radiant energy, photosynthesis.

4.8. MAIN SYMBOLS

In accordance with the previous chapters the symbols are listed under the particular sections where they are first introduced. They are grouped into three broad categories; viz. independent variables, dependent variables and parameters. The groupings are not rigorous: for example, both the specific growth rates are strictly state dependent variables but their symbols are listed together with the state independent parameters.

Symbol	Meaning	Unit
Section 4.2.		
t	time variable	s
Z	daylength	s
I	incident light-flux integral	$J\ m^{-2}\ s^{-1}$
C	ambient carbon dioxide concentration	$g\ m^{-3}$
W	total plant dry weight	g
T	shoot dry weight	g
R	root dry weight	g
F	dry weight of reproductive parts	g
L	total plant leaf area	m^2
σ	specific activity	$g\ g^{-1}\ s^{-1}$

Subscripts C and N denote the specific activities of shoots to carbon uptake and roots to nitrogen uptake respectively.

f_M	incremental composition of plant dry matter with respect to the element M	
μ	specific growth rate	$g\ g^{-1}\ s^{-1}$
Y	yield factor accounting for respiratory loss of carbon by the plant	
α	leaf light utilization efficiency	$g\ J^{-1}$
τ	leaf carboxylation efficiency	$m\ s^{-1}$

Section 4.3		
S	daily radiation integral	$MJ\ m^{-2}$
W	total crop dry matter	$g\ m^{-2}$
L	leaf area index	$m^2\ m^{-2}$
R_C	crop 'dark' respiration rate	$g\ m^{-2}\ s^{-1}$
∇_R	daily respiratory integral	$g\ m^{-2}$
∇_F	daily photosynthetic integral	$g\ m^{-2}$
α_m	leaf maximum radiation utilization efficiency	$g\ J^{-1}$
β	leaf photorespiratory constant	$g\ m^{-2}\ s^{-1}$
m	leaf radiation transmission coefficient	
K	canopy radiation transmission coefficient	

a	constant	$m^2 \ g^{-1}$
b	constant	$m^2 \ m^{-2}$

Section 4.4

F_C	'closed crop' net photosynthetic rate	$g \ m^{-2} \ s^{-1}$
∇_{MAX}	daily photosynthetic integral of a 'closed crop' intercepting all the incident radiation	$g \ m^{-2}$
τ_0	leaf carboxylation efficiency of an upper unshaded leaf in a 'closed crop' canopy	$m \ s^{-1}$

Section 4.5

W	total crop dry matter	$g \ m^{-2}$

Subscripts L, S, R and F refer to leaf, stem, root and reproductive pai weights respectively.

t_f	nominal flowering time (as days after sowing or crop emergence)	d

s_A	specific leaf area	$m^2 \ g^{-1}$
η	partition coefficient	

Subscripts L, S and R refer to leaf, shoot and root parts respectively.

β	proportionality constant for partition of assimilate to the reproductive parts	
γ	leaf senescence/abscission constant	$m^2 \ g^{-1}$
μ	specific growth rate	$g \ g^{-1} \ s^{-1}$

Section 4.6

a_L	leaf growth constant	$g \ m^{-2} \ d^{-1}$
a_S	stem growth constant	$g \ m^{-2} \ d^{-1}$
a_R	root growth constant	$g \ m^{-2} \ d^{-1}$
a_T	shoot growth constant	$g \ m^{-2} \ d^{-1}$
b	the product (Ks_A)	$m^2 \ g^{-1}$
q	shoot senescence/abscission constant	d^{-1}

Section 4.7

t_m	nominal grain maturity date	
t_a	nominal anthesis date	
t_e	nominal emergence date	
H	average daily air temperature	$°C$

$$\left.\begin{matrix} a_0 \\ a_1 \\ a_2 \end{matrix}\right\}$$ constants relating s_A to S and H (eqn (4.111))

$g\ m^{-2}$
$g\ J^{-1}$
$g\ m^{-2}\ (^{\circ}C)$

τ_0^* leaf carboxylation constant $m^3\ s^{-1}\ g^{-1}$
F_{MAX} rate of light-saturated leaf net photosynthesis $g\ m^{-2}\ s^{-1}$

4.9. SUGGESTED FURTHER READING

Section 4.2

Charles-Edwards, D. A. (1976). *Ann. Bot.* **40**, 767–772.
Thornley, J. H. M. (1977). *In* 'Integration of Activity in the Higher Plant' (D. H. Jennings, ed.). The University Press, Cambridge.

Section 4.3

Charles-Edwards, D. A. and Acock, B. (1977). *Ann. Bot.* **41**, 49–58.

Section 4.4

De Vries, D. A. (1955). *Meded. Landb. -Hoogesch. Wageningen* **55**, 277–304.
Monteith, J. L. (1965). *Ann. Bot.*, **29**, 17–37.

Section 4.5

Charles-Edwards, D. A. and Fisher, M. J. (1980). *Ann. Bot.* **46**, 413–423.

Section 4.6

Charles-Edwards, D. A. and Fisher, M. J. (1980). *Ann. Bot.* **46**, 413–423.
Charles-Edwards, D. A., Fisher, M. J., and Campbell, N. A. (1980). *Ann. Bot.* **46**, 425–434.

Section 4.7

Hunt, R. (1978). 'Plant Growth Analysis', Studies in Biology, Vol. 98. Edward Arnold, London.
Austin, R. B. (1978). *ADAS Quarterly Review* **29**, 76–87.
Austin, R. B. (1981). *In* 'Opportunities for Increasing Crop Yields' (R. G. Hurd, P. V. Biscoe and C. Dennis, eds). Pitman, London.